AGRICULTURAL EXTENSION SYSTEMS
ISSUES AND APPROACHES

By the same author

- Extension Education for Human Resource Development.
- Modernising Indian Agriculture in 21st century: Challenges, Opportunities and Stategies.
- Agribusiness and Extension Management.

About the Editors

B.S. Hansra, a Ph.D in Educational Communication from the Ohio State University, Columbus, Ohio (U.S.A.), is currently Director, School of Agriculture, Indira Gandhi National Open University, New Delhi. Earlier, he was an Assistant Director General (Agri. Extension), Indian Council of Agricultural Research.

Dr. Hansra has 37 years of teaching, research and extension experience. He has to his credit 16 books and more than 200 research and popular articles. He has been given many awards and honours, including Mendenhall Memorial Award and Fellow of Indian Society of Extension Education. Also, he has been the Vice-President of International Extension Forum.

P. T. Suraj is Assistant Professor (Livestock Production) at the Kerala Agricultural University. His areas of interest include Smallholder Integrated Farming, Agricultural Policies and Multi-disciplinary Agricultural Development. He has presented a good number of scientific papers at various fora.

P. N. Ananth is Assistant Professor of Rural Development at Jimma University, Ethopia. As a developmental researcher, he has worked on project funded by World Bank, United Kingdom Government, DFID, Cornell USA, World Fish Centre in Malaysia, Union Ministry of Agriculture, and Indian Council of Agricultural Research, New Delhi.

The International Society of Environmental Protection along with the Zoological Society of India gave him the Young Scientist Award for his paper on 'Fisheries Extension and Agricultural Extension Systems'. Dr. Ananth has four books, 25 research papers, five popular articles, three extension brochures and a manual to his credit.

M. J. Chandre Gowda is a Sr. Scientist (Agri. Extension), Zonal Coordinating Unit of ICAR, Bangalore (Karnataka). He has worked in various capacities at UAS, Bangalore; Canara Bank and Indian Institute of Horticultural Research.

Dr. Gowda's areas of specialization include Innovative Extension Approaches, Participatory Rural Appraisal, Participatory Impact, Monitoring and Evaluation and Rural Development through Vocational Education. He has to his credit two books, 23 research papers and 19 popular articles.

AGRICULTURAL EXTENSION SYSTEMS
ISSUES AND APPROACHES

Edited by
B.S. Hansra
P.T. Suraj
P.N. Ananth
M.J. Chandre Gowda

CONCEPT PUBLISHING COMPANY, NEW DELHI-110059

Cataloging in Publication Data—DK
Courtesy: D.K. Agencies (P) Ltd. <docinfo@dkagencies.com>

National Seminar on Alternative Extension Approaches (2004 : Vellanad, India)
Agricultural extension systems : issues and approaches / edited by B.S. Hansra ... [et al.].
p. cm.
Papers presented at the National Seminar on Alternative Extension Approaches, held at Vellanad on 22nd February 2004.
Includes bibliographical references.
Includes index.
ISBN 13: 9788180695025
ISBN 10: 8180695026

1. Agricultural extension work—India—Congresses. 2. Agriculture—Technology transfer—India—Congresses. I. Hansra, B. S. (Baljit Singh), 1945- II. Title.

DDC 630.715 22

ISBN-13: 978-81-8069-502-5
ISBN-10: 81-8069-502-6

First Published 2008

Published and Printed by
Ashok Kumar Mittal
Concept Publishing Company
A/15-16, Commercial Block, Mohan Garden
New Delhi-110059 (India)

Phones : 25351460, 25351794
Fax : 091-11-25357103
Email : publishing@conceptpub.com

Foreword

One of the greatest challenges in Indian agriculture is filling the growing gap between scientific know-how and field level do-how. The gap between potential and actual yields is very wide even with the technologies on the shelf, as observed from the National Demonstrations data of the ICAR. The average productivity of most of our farming systems is low. Thus, there is a large untapped reservoir which can be availed of through an integrated package of technology, extension, input supply and assured and remunerative arrangements.

The Eleventh Five Year Plan document refers to the prevailing knowledge deficit in rural India. The present book on Agriculture Extension Systems provides a road map for overcoming the knowledge deficit. The book is comprehensive and deals with both public and private sector initiatives in extension. I congratulate Dr. B. S. Hansra and all the editors of this timely publication. I hope it will be widely read and used in order to help our farm women and men to improve the productivity, profitability and sustainbility of the major farming systems of our country.

M.S. Swaminathan

Preface

The world over Agriculture Extension is in focus due to the new challenges in technology transfer and sensitivity in government policies towards funding public extension. It is high time that we search for good extension approaches adoptable for replication. The approaches of many public extension services, private and non-government sectors are worth mentioning for replicating at different agro climatic zones. The book contains articles on different aspects of developments in extension and the efforts taken by different authors in their respective areas/work. A good approach in agricultural extension needs wide dissemination. The book is a synthesis of the presentations by these contributors. They deserve sincere thanks and appriciation.

We thank everyone who helped directly or indirectly for realization of this book. We express our gratitude to the Concept Publishing Company, New Delhi also for accepting and publishing this book.

Editors

CONTENTS

Section - II

EXTENSION APPROACHES

Introduction

This book on 'Agricultural Extension Systems: Issues and Approaches' is the outcome of the National Seminar on Alternative Extension Approaches held at Mitraniketan, Kerala. More than 200 research scholars, farmers, NGOs, extension scientists and extension personnel participated and deliberated upon the subject for two days. During the deliberation all were concerned that many developmental and research organizations that strive hard to transfer their innovations to their client groups fall short due to lack of innovative approaches in technology transfer. On the other hand some have excelled in transferring of technologies because of their innovative extension approaches. During the seminar many innovative extension systems and apporaches were presented and discussed. It was decided that the important and relevant systems and approaches be documented in the form of book. Accordingly, twent-two papers which were relevant and important were identified.

The book has been divided into two sections : Section One deals with different extension systems, whereas the Section Two enlists various innovative extension approaches. There are eight chapters in Section One. In Chapter 1, an attempt has been made to highlight the major concerns about the implementation of different extension systems under different conditions in Kerala state. The possibilities for desired transformation from the present scenario are discussed. In Chapter 2, points like continuing marginalization of farmers, low growth rate in agriculture and poor income generating capacity that are responsible for reshaping of the extension system are dealt with. New role of agricultural extension in shifting farming activity from traditional, scientific and commercial methods towards market orientation and value addition of crops has been emphasized in Chapter 3. Livestock owners have vested much hope in the Panchayati Raj, and this new developemental model has been stressed in Chapter 4. The Chapters 5, 6, and 7 refer to private extension services provided by an individual or organisation other than the government. Such set-ups are providing single service or integrated services to the farmers. It has been emphasized that private extension will make extension services more decentralized, farmer participatory, flexible and provide ownership rights to the clients. The last Chapter of Section One reviews the public research and extension that played a vital role in bringing about the green revolution. In this context, the extension focuses important challenges in the area of capacity building, relevance and accountability.

In Section Two, different extension approaches have been discussed. Chapter 9 dealts with Self Help Groups — an effective extension strategy for rural development. The Self Help Groups have been recognized today as the pillar for poverty alleviation programmes of this century. As micro credit is widely appreciated in Chapter 10, importance of organizing marketing efforts, technology upgradation and diversification for groups associated with the programme have been discussed in Chapter 11. It has been often emphasized that Cyber Extension would be the major form of technology dissemination in the near future. The ongoing IT revolution has opened up tremendous opportunities for providing access to information as well as interactive distance learning. All these aspects have been reviewed in Chapter 12. The Chapter 13 gives the overview of Raitha Samparka Kendra — an alternative demand driven extension approach in Kerala. In Chapter 14, the integrated farming system — an innovative and noteworthy extension approach has been analysed. The characteristics of the Integrated Farming System Demonstration have been dealt in detail. Another approach of adaptability and proliferation of techno-interventions under Institutions — Village Linkage Programme of the coastal village in Kerala has been discussed in detail in Chapter 15. Another approach, the Sayahnam in Kerala, is discussed in detail in Chapter 16. This method imparts better understanding to the farmers about the common field problems and their remedies. In Chapter 17, impact of different extension approaches used in

KVK system, especially in Sardar Shahar, Rajasthan has been elaborated. It is emphasised that suitable extension approaches be adopted according to the priorities and objectives of the programme and needs of the future. In Chapter 18, a KVK success relating to women self help groups in the field of women emancipation and empowerment had been illustrated.

Agriculture progress will form the pillar of economic growth and political strength of the country. There is a need to restructure research strategies and extension services to make things farmer friendly. It should also become capable of converting generic into location specific knowledge essential for taking to precision farming. The rural knowledge centres should provide computer aided and Internet connected information services, so that families have timely and relevant meteorological management and market information; this was highlighted in Chapter 19. Chapter 20 has reviewed the extension approaches like farmers discussion groups, field demonstrations and training of KVK, Gandhigram, Tamil Nadu. Chapter 21 details the concept of Krishi Vigyan Mandal, a new approach of farmers to extension in KVK, Jalna, Maharashtra. The last Chapter 22 deals with the extension approach for sustainable aquaculture in India. Details of integrating participatory and group approaches for technology transfer towards sustainable fisheries was also outlined.

These approaches are innovative and would offer deep insight to extension development for effective agricultural growth. This book will prove to be highly useful to people working on technology transfer and training at various levels.

Editors

Section - I

Extension Systems

Chapter - 1

Concerns and Possibilities of Extension System

A case of Kerala

R.M. Prasad

Introduction

An analysis of the present extension system reveals that there are various concerns related to the system, which have to be critically reflected and attended to by the policy-makers. Similarly, an overview of both the mainstream and isolated extension programmes in Kerala highlight the potentials and possibilities for an effective extension service, which could in fact be role model for the national extension system.

An attempt is made in this chapter to highlight the major concerns of the extension system. The possibilities for desired transformation from the present scenario are also discussed.

Major Concerns

Professional Concerns

The present public extension system in the State to a large extent lacks professionalism. Extension personnel in the state do not find space for utilizing their technical knowledge and skills, since majority of their time is spent for mere delivery of inputs and related office administration, which is not 'professional'. Many of the extension personnel in the State had opined that T&V system of agricultural extension implemented earlier was more successful in utilizing and enhancing their professional competencies. Extension service, some how has been wrongly conceived as one of 'providing' supplies and services, as one of 'giving' something to farmers.

However, this supply driven agricultural extension has proven unsuccessful and also unsustainable. This should necessarily be modified as demand-driven and farmer accountable. The major role of competency building of farmers through facilitation and empowerment is gradually getting diffused and in order to overcome this, a new professionalism has to be developed, which should stress on the following principles :

(a) The 'directive' approach of extension should be replaced with 'participatory' approach.
(b) The interventions by the extension system should not be 'authoritative', but 'facilitative'.
(c) The thrust by the extension system should not be on 'adoption' of farm practices, but on 'adaptation' to suit the various agro-climatic conditions.
(d) The organisation of extension service should range from primary production to entire food chains.
(e) The extension service must be reoriented from public to public-private partnership (PPP) mode, including community driven development (CDD).
(f) The extension service must shift from narrow agricultural focus to broader policy contexts, including global impacts.

Thus extension service should be made more participatory and better opportunities for a more interactive process of technology development must be created (Pretty and Chambers, 1993). In this context, group approaches having a broader professional framework in the form of 'Training for Transformation' (TFT) rather than Transfer of Technology (TOT) have to be adopted. Extension has to focus on processes related to synergy, complementarity and integration rather than paternalism, control and specialisation, which is the order of the day.

Organisational and Management Concerns

The public extension system alone will not be able to meet the increasing information needs of farmers. The government also cannot fund public extension services to a large extent due to financial constraints. This has compelled the emergence of other alternate extension approaches and strategies. Private Extension Service (PES) is thus emerging as an alternative to the public funded extension system. However, private extension will not and cannot replace public extension *in toto*. Hence the public extension system has to be totally reorganized by providing ample opportunities and space for private extension services to operate. Based on the services, which they can provide, the services could be grouped into three categories as replacement services, supplementary services and complementary services. The public extension system may focus on facilitating private extension agencies to develop necessary competencies to undertake various responsibilities and also properly monitor and regulate their actions.

Technological innovations (TI) and Human Resource Development (HRD) are two important dimensions of effective extension service. While public extension system will be more effective in providing technological services, private extension system will be more effective in organizing and mobilisng farmers. Hence the extension services of the public system if supplemented by the private sector will be more effective.

The public extension system should realize that the need of the private sector is not for the technological skills *per se*, but for their social skills combined with the strategic resources and infrastructure, which they can provide. The concept of adaptive management strategy put forth by Lee (1999) has relevance and application in the management of extension services. The public extension system should try to formulate and implement an adaptive collaborative management style in which more partners from private and NGO sectors are integrated with the public system for achieving efficiency and sustainability.

Demographic Concerns

Homestead farming has been one of the survival strategies of the farmers of Kerala since time immemorial. The statistics on the operational holding of Kerala reveal that though more than 92 per cent of the farmers in Kerala are marginal farmers (with land below one hectare), they occupy an area of only less than 50 per cent of the total cultivated area in the State. A clear distinction exists between homestead farmers and plantation agriculturists, which gets reflected in their differential needs relating to extension services also.

How far the public extension system could cater to the needs of farmers in Kerala still remains a big question. While homestead farmers are forced to depend on the public extension system, plantation farmers can afford to obtain paid extension and consultancy services. Even then the services of the public extension system are to a large extent utilized by the commercial farmers, rather than by the subsistence-oriented farmers. In this context, different extension strategies to fit various farm types are suggested, as put forth by World Bank (2003):

(a) *Commercial farmers*—who produce entirely for the market. Such farmers prosper where there is a market-friendly enabling environment for private investment and initiative.
(b) *Small producers on family farms*—who are linked to markets, but have limited assets. With effective extension strategies, many of the farmers can build their asset base and make the transition to commercially oriented farming.
(c) *Subsistence oriented farmers*—who operate in less-favoured production environments and lack most types of assets. They have varied livelihood strategies; often operate outside the market. Many of these farmers are also located in marginal areas with limited infrastructure and few options for agricultural improvement. The public extension system should strive to provide free extension services to small and marginal farmers operating in less favourerd production environments, whereas extension services could be charged in the case of large and commercial farmers.

Technological Concerns

The State extension system provides direction and guidance towards technological empowerment of farmers, which has to be further strengthened. There are many technologies available which are not yet known to the potential users. Similarly, technologies based on improvement of traditional technologies (upscaling) and based on modifications of modern technologies (downscaling) are also available. The aim of extension service should be not only to communicate the available technologies to farmers, but also to provide opportunities for technological testing and adaptations so as to redefine and fine-tune the technologies to meet the needs of the farmers.

The Report of the Commission on WTO concerns in agriculture, Government of Kerala (2002) has rightly pointed out that the state extension system should take steps to shift to an era of precision farming in order to maximize factor productivity and minimize cost of cultivation. The Commission has recommended the following measures for empowerment of farmers :

(a) *Productivity enhancement*—by bridging the existing gap between potential and actual yield with technologies on the shelf.
(b) *Quality improvement*—through a quality literacy movement for producers and consumers, and by strengthening the infrastructure for sanitary and phytosanitary measures for both domestic and export markets.
(c) *Profitability enhancement*—through concurrent attention to production efficiency and higher factor productivity, as well as post-harvest technology, agro processing, etc.
(d) *Sustainability improvement*—through attention to ecological foundations essential for improving productivity in the long-run.
(e) *Stability of production and income*—through proactive advice on trends in home and external markets and measures like market stabilization fund and insurance.

Economic and Ecological Concerns

An important feature of the scientific technology in the field of agriculture is that it regards nature as a set of 'resources' divorced from man to be 'exploited' by him in the process of making profit. One area where this conflict has come up most sharply is in the problem of common pool resources. The present development thrust through the use of technologies has led to either privatization or alienation of the 'commons' leading to problems not only to the access to and use of, but also to the maintenance and survival of these resource bases.

A visible impact of technological change in agriculture is the high dependence on external inputs, which has put farming under the control of MNCs, who are known for exploiting natural resources and have been responsible for several tragic environmental disasters. Still another problem associated with agricultural technologies is the threat to bio-diversity. Extension system has to seriously think about the impact of agricultural technologies on sustainable development.

Technologies in farming should be linked with man and nature, to the advantage of both. The need for 'resource conserving' technologies in farming must be emphasized and disseminated to the farmers. Visvanathan and Parmar (2001) had indicated that as Swaminathan had opined sustainability is a multi-dimensional concept with three critically interacting dimensions—ecology, economics and equity. The underlying philosophy is that in order to be sustainable, development should be environmentally compatible, economically efficient and socially equitable.

It is noticed that livelihoods are often viewed much more than cultivation, which are more varied. As Chamber (2002) had suggested, the degree to which farmers complicate and diversify their livelihood strategies can be illustrated in six dimensions, viz., capability, person and activity, enterprise, social relationships, season and interlinkages. Extension services should focus on these dimensions of farmers and the understanding about the complexity and diversity of the various dimensions of livelihood will enable the extension personnel to evolve suitable extension programmes, which will be sustainable.

Considering the concerns discussed above, extension service in the State should be reorganized in the following lines:

(a) A uniform extension strategy for farmers with varying farm types, production systems and market opportunities will not be effective. A different extension strategy for each major farm type, viz., commercial farmers, small producers on family farms and subsistence oriented farmers has to be formulated. To link farmers to input and technology providers, strong producer organisations and community driven groups should be formed and made use of.

(b) The intervention by extension workers should not be authoritative, but facilitative. The appropriateness of technology and technological niche may be considered by the extension workers. The thrust by extension workers should not be on adoption of technology, but on adaptation to suit farmers' situations. Adaptive collaborative management strategy should be the functional paradigm to be exercised by the extension workers.

(c) The information needs of different farmer categories call for a clear conceptualisation of the roles of extension workers. The emerging trends in agriculture warrant a more realistic and rationally oriented professional approach in terms of farmers' needs and requirements. The extension system should play a more pro-active role in providing an enabling environment for farmers to act.

(d) The pre-packaged information (tell me what to do) for farmers should be discouraged. Farmers want guidance and need based information to solve their varied problems. Technology packages have to be evolved for the resource system on a participatory mode. Participatory Technology Development (PTD) must be given due emphasis and prominence by the extension system.

(e) Communication by extension personnel which hither to had followed the DAD model (decide, announce, defend) should be replaced by CCC model (convey, consult, convince). Considerable conceptual skills have to be developed in the farmers to enable them to take forward looking decisions.

(f) Markets are now the driving force for agricultural growth.

Extension should focus on strengthening farmer to market linkages and should address this through supply chain management structures, improving product quality, establishing appropriate post-harvest systems and assisting the farmers in refining methods in product certification and improving food safety.

(g) Public extension system should reorient towards public private partnership (PPP) mode, including community driven development (CDD). CDD offers the potential for increased community participation and effective decision making. More partners from private and NGO sectors may be integrated with the public extension system for achieving efficiency and sustainability.

(h) Extension service should focus on extending the gains of research to small and marginal farmers by creating an enabling environment in which functional mechanisms are established for providing value added information and services to the farmers. The extension system should try to scale up good practices and innovative experiences in the field so as to reach all categories for farmers. This should be an integral part of the extension system and methods for capturing, validating, disseminating and adapting good practices have to be evolved by the extension workers.

(i) The extension system should focus on agricultural diversification, which should be viewed as the main vehicle for rural employment generation and income argumentation, especially by connecting inter-enterprises and inter-sectoral linkages. The focus on promotion and development of small and medium enterprises (SME) enable the farmers to secure sustainable employment and income. This could be achieved through translating the concept of 'convergence' into action in the field.

(j) Most of the policy measures currently adopted for supporting agriculture unfortunately turn to be powerful disincentives against sustainability. A good example is the case of subsidies for farmers. Farmers are interested in receiving give-aways and they evaluate programmes in advance on the basis of what give-aways they are offered rather than on merits. It is necessary that a relook be attempted on this and such related issues.

REFERENCES

Chamber, R. (2000) *Whose reality counts? Putting the first last.* ITDC Publishing, London.

Government of Kerala (2002) Building on sustainable agricultural trade security system for Kerala—Report of the Commission on WTO Concerns in Agriculture, Government of Kerala, Trivandrum.

Lee, K.N. (1999) Appraising adaptive management. *Conservation Ecology* 3 (2) : 3.

Pretty, J.N. and Chamber, R. (1993) Towards a learning paradigm. New professionalism and institutions for agriculture. Discussion. Paper No. 334, IDS, University of Sussex, Brighton.

Vaidyanathan, A. (2001) Poverty and development policy. *Economic and Political Weekly* 36 (21): 1807–1822.

Visvanathan, S. and Parmar, C. (2001) A bio-technology story — Notes from India. *Economic and Political Weekly* 37 (27): 2714–2724.

World Bank (2003) *Reaching the rural poor – A renewed strategy for rural development,* Washington DC.

Chapter - 2

Reshaping of Extension System
Potentials and Prospects

Jayasree Krishnankutty and *S. Parvathy*

Introduction

The extension system was purely public sector-based in post-Independence India. The role of extension was considered as transfer of technology, predominantly and the focus was the farmers, relying on an inductive approach, mostly. The system has undergone gradual changes over the decades with major approaches like T & V, group approaches and area approaches to a stage where it finally acknowledges the potential of private sector to deliver extension. But whatever is the approach it remains a serious question whether the system has succeeded to serve, which is accentuated by facts like continuing marginalisation of farmers, low growth rate in the agricultural sector and poor income generating capacity of agriculture. All these point towards the need for modifying the extension system.

Extension is meant to act as the bridge between the farming system and the research system. If this disparity is to be kept minimum, the stakeholders in agriculture have to be always kept abrest of the technological advancements. This forms the major role of extension of technology dissemination. But if technology generated has to be need based an equally strong mechanism should be there to bring back the responses from the stakeholders. What apparatus is there with the extension system to meet this requirement is to be seriously examined.

Content of Extension

Content of extension meant mainly the transfer of technology methods social science aspects to explain understand and improve the clientele's behaviour, management aspects, programme implementation and evaluation methods. This content has to expand to fit in with the changing needs, problems and challenges experienced by the agricultural scenario. Extension should include elements of soil and water conservation, bio-diversity, diversification to Horticulture, value added products, development of rural infrastructure and agro-enterprises and creation of employment in rural areas (Department of Agricultural and Co-operation, 2002).

Extension Mechanisms

The extension machinery is supposed to operate in a two-way process. Information dissemination to the clientele and bringing back the needs, field level difficulties, and location/cultural/social oriented variations back to the research system. It is the latter part that seems not to be stressed mostly: While clear-cut, experience based and much researched mechanisms are there for technology dissemination, an equally strong and well defined feedback mechanism seems to be lacking. A permanent clientele research cell to suit various situations will ideally serve this purpose. To gather, integrate and access information according to need, a repository maintenance mechanism should form part of the extension system. This will help to integrate information from various centres and design client/system specific packages.

Survey cell-mechanism for integrating information from various centres and formulating a comprehensive central package.

Computer based technology is the biggest factor for making change in the extension mechanism. IT brings new information services to rural areas over which farmers, as users will have much greater control over the current information channel. Even if every farmer does not have a computer terminal these could become readily available at local information resource centres, with computers carrying expert systems to help farmers to make decision. It will help in bringing together the expertise of farmers and the best available scientific knowledge to develop farms and local agricultural economics.

Extension Approach

Extension should have a multifaceted approach. It should serve the individual without losing view of the group interest and the greater common objectives in general. This calls for a host of approaches, new and modified.

Farming system approach, watershed approach participatory extension approaches can be some among them.

The various approaches developed earlier were sometimes off target or they were sometimes less relevant to the objectives and socio-economic circumstances of small land holder farmers or were in appropriate to the agro-ecological conditions.

The farming system research methodology is centred in problem-solving, and is system oriented, interdisciplinary farmer oriented and interactive. It emphasizes the role of constant diagnosis and on-farm trials as a way of facilitating linkages among the farmers, researches and extension workers. Farming system perspective also involves for greater concern for local resources utilization and use of traditional knowledge.

Watershed Approach

Watershed approaches can be planned and executed in five steps such as, recognition phase (Analysis of the cause of the problem, Development of alternative solutions of the problem), restoration phase (Selection of the best solution), application of the solution of the problem to the land, protection phase and improvement phase.

Participatory Approach

The entire process in which extension personnel facilitate the generation and dissemination of agricultural innovations together with farmers.

Project Based Extension

In which a group of farmers can work on a problem or project like piggery unit and vermicomposting, kudumbasree units for women.

Individual Farmer Approach

Focuses on building up relations, other than mere transfer of technology. Help in more grassroots contacts, more aware of farmers' needs, effective communication, specific interest can be brought into limelight.

The trend of invisibility, non-recognition and unaccounting of women in agriculture is still continuing. Making self-help groups of women for products and marketing will be a solution to this situation.

Extension Roles

The roles of extension should expand to include linkage, co-ordination, integration, pooling of information and repository maintenance.

Collaboration and networking enable organisation to extend and thereby improve their outreach abilities in order to serve as many target groups and individual beneficiaries as possible. Extension needs to take up increasingly the role of enabler, through its role as a doer should also continue but in a qualitatively different way. The field step could be the identification of the different elements in the agricultural knowledge and information system and redesigning this system in a way that information flow among these elements is improved (Sulaiman and Hall, 2002). Extension can oversee this process and should be always on the look out for adding new aspects to the system.

Extension Agencies

It is now widely agreed upon that public sector extension has to be supplemented with private extension. Small scale agri-business units, agricultural firms, agricultural professional groups, paratechnicians, progressive farmers and farmers' organisations/ co-operatives are agencies which engage in extension (Chandrasekhara, 2001). Besides these, a large number of NGOs and self-help groups also provide private extension. The potential of private extension has to be properly tapped, promoted, their efforts integrated and linked with public extension.

Conclusion

The concept and methodologies of agricultural extension have to reshape themselves to suit the changes in agricultural situation. Extension has to include broader content, devise a host of new approaches and assume more diversified roles so as to equip the stakeholders in agriculture to face the challenges in their field. If the role of extension as the bridge between the farming system and research system has to be properly carried out, the feedback mechanism available with the extension system has to be strengthened and modified. These may call for some overhauling in the system, but it will definitely help the extension to become true ambassadors of agricultural development.

REFERENCES

Chandra Sekhar, P. (2001). Private extension: Indian way. (Chandra Sekhar, P. (ed.). *Private Extension - Indian Experience,* National Institute of Agrl. Extension Management, Hyderabad.

Department of Agriculture and Co-operation (2002). *National Agricultural Policy. http://agricoop.nic.in/agpolicy02/htm.*

Sulaiman, V.R. and Hall, A.J. (2002). *Beyond Technology Dissemination - Can Indian Agricultural Extension Reinvent Itself?* Policy Brief 16, New Delhi: National Centre for Agrl. Economics & Policy Research.

World Bank (1983). *World Development Report 1983.* New York. Oxford University Press.

Chapter - 3

New Role of Agricultural Extension to Face the Present Challenges for Effective Transfer of Technology

K. Anand Singh

Introduction

Indian Agriculture is undergoing a rapid transformation under globalization from a simple making a living to a complex endeavour. The focus of agriculture is increasingly shifting from raising the crops by traditional to more scientific and commercial methods and the concept is changing towards market oriented and value addition crops.

The extension system (Department or University) had played a key role in dissemination of agricultural technology to farming community and with the result there had been enhanced crop productivity. But despite the efforts the latest technology generated by research system may be with regard to new varieties, improved production technology, market awareness is not reaching the ultimate users—farmers, to the expected levels especially resource poor due to varied operative and administrative constraints. It is heartening to read from newspapers that farmers are frustrated, depressed and resorting to suicides and extension is one to be blamed besides other systems for these mishappenings. The role of extension functionaries today is more challenging and they have to equip themselves to cope up the drastic changes and dire needs.

Some of the problems faced by the farming community today are slow pace of technology adoption as per the changing scenario, lack of availability of improved Agricultural inputs, lack of Market information for different Agri-produce, poor capital and credit support, lack of inputs and resources, poor participation of women as stakeholder, lack of coordination and organization among farming communities, poor and untimely help from Government agencies, Lack of enterpreneurial skill in agri-based occupations, lack of managemental technology for optimum resource utilization.

Considering the globalization, liberalization and introduction of WTO there is tremendous diversification observed in farmer's expectation from extension system. It is very difficult to fulfil as these expectations are great challenges for extension system. Some of the new roles of Agricultural Extension in responding to challenges are:

1. Identify the potentials of *agribusiness* and its relevance to improve farming situation by raising market oriented and value addition crops.
2. Providing *need-based skill oriented vocational training* back up to farmers, farm women and rural youth.
3. Conducting frequent *exposure visits* of farmers to successful farms in the districts and also other states.
4. Adequate *training of extension personnel* by updating their knowledge on changing concepts—technology, market etc., and preparing them to be more pro-active particularly when there is pest/disease or any disaster/calamity.
5. Women carryout more than 70 per cent of operational activities in agriculture and allied sectors. But her participation in developmental activities is not even 5 per cent. Women in rural societies find difficult to obtain raw materials, resources, marketing network, access to credit, agricultural education and employment as reported by large number of research and evaluation studies. *Women empowerment* is a result of participation in decision making. Women should be involved in every stage, like technology development, technology dissemination, technology utilization and policy formulations in agricultural extension.
6. *Cyber Extension*: Access to information and improved communication is a crucial requirement for agricultural development. The on-going IT revolution has opened up large

opportunities for providing access to information as well as to interactive distance learning in rural India. The applications of IT in extension education need to be experimented. Access to information is not only fast but also much cheaper than the traditional methods. Even if every farmer does not have a computer terminal, these would become rapidly available at local information resource centres, with computers carrying expert systems to help farmers to make decisions. The agricultural extension machinery should be totally revamped to meet the information hungry farmers of tomorrow. Computer literacy in agricultural universities and other educational institutions should be brought in on a war footing.

7. There is a need to have *mobile extension* service to diagnose field problems and understand advisory programmes, to identify farmers training needs, extension scientists with efficient communication skill and incentives to extension scientists.
8. Based on the information needs of farmers regarding various aspects of farming through telephone more emphasis should be given to make *'Help-line'* popular in districts in the region and provide free call service to the farmers to solve their farm problems.
9. Extension personnel should also facilitate in conducting *farmers field schools* in the situation where farmers cannot visit Research Stations for training. Training's can be given to farmers at field level in the villages by selecting a good village and field for imparting latest technology adopted.
10. Better communication for *strengthening for linkage* between research extension and farmers.
11. Providing *market information* to farmers.
12. Providing information on *schemes and policies* of Government.
13. Help farmers to establish cottage *agri-based industries* and other enterprises like Poultry, Goatry, Fisheries, Apiary, Sericulture etc.
14. Establishment of *Agricultural Technological Information Centre* (ATIC) to provide the information to farmers regarding various aspects of farming like latest seed varieties, pests and disease identification and method of their control, market

information etc., and complete information pertaining to Agriculture and allied aspects.

15. Establishment of *farmers clubs and self-help groups* consisting of separately for men and women and providing technological back-up by the Krishi Vigyan Kendra through training and other appropriate means, facilitating and motivating them to save money based on the interventions identified. Infuse innovativeness in the group and also facilitate for participatory technology development and use the group for not only income generating activities but also in solving other socio-economic problems of the village and all-round development.

The extension service, therefore, is to function as a part of an interdependent technology development, transfer and utilization system and must achieve a two-way flow of information that is (i) information on farmers needs, problems and priorities and (ii) Information on solutions to farmers problems to enable and to empower them to take correct decisions. If agricultural extension service takes care of these two, perhaps many of the limitations of extension service can be tackled easily encompassing the challenging needs of 21st Century.

Considering the globalization, liberalization and introduction of WTO there is tremendous diversification observed in farmer's expectation from extension system. It is very difficult to fulfil these expectations and are great challenges for extension system. The major expectation of our farmers from extension system is provision of appropriate technology as per their own farming situation, necessary arrangement for supply of needed inputs, on spot solution to problem and credit delivery imparting required skills, marketing and processing facilities and schedule of visit of extension personnel to be followed are the major diversification of farmers expectation. These diversifications of expectations are considered as great challenges for Agricultural Extension. For this it requires broader role of agricultural extension and it is essential to have strong and efficient agricultural extension system for 21st century, i.e. 2020.

Chapter - 4

Facilitating and Restraining Forces in Carrying out Extension Activities Under Panchayati Raj

A Force Field Analysis

Shubhangi Baburao Tajane and *P.J. Rajkamal*

Introduction

The study was conducted among the veterinary surgeons of the Animal Husbandry Department (AHD) and those working in the Thrissur district of Kerala. These veterinary surgeons as part of their roles perform extension work under Panchayati Raj. The objectives of the study were identifying those forces or factors in the work environment facilitating as well as that restraining extension work under Panchayati Raj and identifying solutions to counter these forces.

Methodology

Force field analysis, the concept put forth by Kurt Lewin *et al.* (1951) and the procedure of carrying out the same as suggested by Pareek *et al.* (1981) with modification, were followed in the present study to determine the facilitating and restraining forces in carrying out extension work under Panchayati Raj. Accordingly different data collecting tools were used, namely questionnaires, interview schedules, and group interviews. Initially questionnaires were used and 103 veterinary surgeons returned them out of total of 142 veterinary surgeons of Animal Husbandry Department (AHD) and were working in Thrissur district to whom questionnaires were sent. From the responses to the questionnaires it was seen whether there were any strong or very strong restraining forces as well as any weak or very weak facilitating forces. For this, responses to initially identified facilitating and restraining forces were obtained in a five-point continuum, namely very strong (≥4.5), strong (≥3.5), somewhat strong (>2.5 to ≤ 3.5), weak (≤ 2.5), and very weak (≤ 1.5). Solutions to those selected forces were identified in-group interviews during monthly conferences of veterinary surgeons. Commonly held views and opinions were consolidated for this purpose. Later, prioritization of the different solutions mooted was done by 30 veterinary surgeons responding to the questionnaires. Interview schedules were used for this purpose. To determine priorities reverse weight age scores such as 3, 2 and 1 were assigned to I, II and III preferences etc. respectively. The total score for each solution for all the 30 respondents was worked out and the number of respondents to attain the mean weight age score divided this total score. This mean weight age score formed the basis for determining the overall priority.

Results and Discussion

As per data in Table 4.1, the only weak facilitating force identified was a lack of enough motivated youth and women participants to animal husbandry development programmes. Though many solutions were mooted to strengthen this force the most important was the need of more viable self-employment projects that could increasingly attract youth and woman into the development process. It is only common knowledge that many of our self-employment projects lack viability because they are non-feasible physically and non-enduring economically. Technological intangibility has been yet another impediment.

There were many strong forces that could restrain extension activities. Delayed clearance of projects was one and the solution to which first priority was assigned timely allocation of funds. Delayed funding of projects itself was pointed out as another strong restraining force. The first and foremost solution voiced was doing away with the present delay in treasury transaction. Treasury ban has been frequent and the situation should improve for the better, definitely through efficient financial management.

Delayed beneficiary identification and defective identification of beneficiary were two other strong restraining forces identified. Gramsabha is the competent authority to identify beneficiaries.

Table 4.1 : Facilitating and restraining forces and the solutions

Sl. No.	*Forces (Factors)*	*Mean score*	*Solutions (first priority)*	*Mean score*
	A. Facilitating force(weak force)			
1.	Lack of enough youth and women participants	2.49	More of viable self-employment programmes for youth and women	3.13
	B. Restraining forces (strong forces)			
2.	Delayed beneficiary identification	3.94	More publicity for Gramsabha	2.70
	Delayed funding			
3.	Inadequate follow up	3.75	Avoid delay in treasury transaction	3.50
4.	Defective identification of beneficiaries	3.74	Comprehensive projects needed that is market linked and lasting projects	2.13
5.	Resources are not made available to people at right time	3.80	Beneficiary identification should be based on felt needs of people Timely allocation of funds	2.86
6.	No proper arrangement for training the development staff	3.67	A befitting training curriculum that address the needs of the trainees	3.00
7.	Inadequate monitoring	3.64	Monitoring should be done in a phased manner	3.30
8.	Inadequate evaluation	3.63	Technical personnel should be included in the evaluation committee	3.10
9.	Panchayat authorities inability in decision-making	3.61	Guiding them technically by qualified people.	2.83
10.	Hasty implementation	3.54	Timely clearance of projects	3.0
11.	Lack of clear direction	3.54	Issuing timely and relevant guidelines orders etc.	1.7
12.	Dual control of the staff	3.54	Proper coordination between departments and local authorities	1.60
13.	Political polarization	3.53	Common development agenda	1.53
14.	Delayed clearance of projects	3.52	Timely allocation of funds	2.30

But notwithstanding the statutory significance of Gramsabha they are not held regularly or fairly. People loose faith in Gramsabhas and they fall short of attendance. Now the whole confidence building process might have to begin again. This could be the reason why the need for more publicity for Gramsabha was highlighted as the most important solution to tide over delayed beneficiary identification. However, regarding defective identification of beneficiaries the solution of selecting beneficiaries on the basis of felt needs was assigned top priority.

Dual control and lack of clear direction were two other strong restraining forces, and the most important solutions suggested respectively were proper coordination between AHD and local PRI, and issuing timely and relevant guidelines, orders etc. AHD and PRI should no more remain as islands, instead should be effectively linked together establishing proper communication.

Lack of proper arrangement of training of the development staff, and panchayat authorities, inability in decision-making were two other strong restraining factors. The foremost solutions suggested respectively were a befitting training curriculum that addresses the needs of trainees and guiding the panchayat authorities by technically qualified people. A job-cum-task analysis of the development or extension functionaries is, therefore, a prerequisite to decide training needs. Sometimes the panchayat authorities lack technical knowledge that the technically qualified extension functionaries must impart. There needs to be a learning culture in the whole system, so that people learn from each other's experience and skills.

It is everybody's experience that usually projects are implemented in haste at the fag end of the financial year. This could be the reason why hasty implementation was cited as a major restraining force. Timely clearance of projects was the wayout suggested in this connection. Resources could not be made available to people at right time was yet another restraining force and the most important solution being timely allocation of funds. In a democracy political opinion tend to be divided. In the present study, political polarization was cited as a strong restraining force and the most important solution proposed was a common development agenda to work together. It is need of the time that irrespective of factionalism people must work together on matters of their development and welfare.

Inadequate monitoring, evaluation, and follow up of programmes were voiced as strong restraining factors. The respective and the foremost solutions suggested were carrying out monitoring in a phased manner, including technical persons in the evaluation committee and need for marked linked comprehensive projects. Even in the new development model, monitoring, evaluation, and follow-up were weak was a matter of serious concern and thought. A development model is not worth if the past mistakes still hold.

Conclusion

All the above issues should draw the immediate attention of the authorities concerned, definitely the lead players, namely Government AHD and PRI. There has to be more discussion and debate on such issues involving all stakeholders. The present state of things must improve anyhow, for the livestock owners have vested much hope in Panchayati Raj, the new developmentel model.

REFERENCES

Kurt Lewin. (1951). *Field theory in social science: selected theoretical papers.* (New York: Harper and Brother.

Pareek U., Rao, T.V. and Pestonjee, D.M. (1981) *Behavioural Process in Organizations.* New Delhi, Oxford and IBH, pp. 262-69.

Chapter - 5

Private Extension and Cyber Extension in India

A.I. Jose

I. PRIVATE EXTENSION

Private extension refers to the extension services provided by an individual or organization other than government who is providing single service or integrated services to the farmers.

Private Extension Providers

Under Indian context, the main categories of private extension providers are :

1. Unemployed graduates in agriculture and allied subjects
2. Agricultural consultants/firms
3. Para-technicians
4. Progressive farmers
5. Farmer organizations/cooperatives
6. Non-governmental organizations (NGOs)
7. Krishi Vigyan Kendras (KVK)
8. Agri-business companies
9. Input dealers
10. Newspapers
11. Agricultural magazines
12. Private television channels
13. Private sector banks
14. Internet
15. Donor agencies

From 35 Agricultural Universities and five deemed universities of the country nearly 12000 graduates in agriculture and allied subjects are produced annually and only about 2000 of them find employment in government establishments. Obviously, the agricultural graduates have to involve in private extension services. The Government of India introduced agri-clinic and agri-business centres to provide alternative self-employment to agricultural graduates under a central sector scheme in collaboration with the SFAC and MANAGE, Hyderabad.

The term 'Para technicians' refers to professionals possessing necessary skill required to perform one or more agricultural activities, but without a professional degree.

A typical example for farmer cooperatives involved in private extension is the Kaira District Cooperative of Milk Producers (Gujarat) which is known under the name AMUL which has a turn over of Rs.500 crores per year covering 9.7 lakh farmers and 150 extension workers. UPASI (United Planters Association of Southern India) is a well-known example of a farmer organization undertaking extension services.

At present there are more than 20,000 registered NGOs in the country involved in rural development and some of them with excellent performance (e.g. Mitra Niketan). Out of 551 ICAR sponsored KVKs in the country, 90 are run by NGOs.

The agri-business companies (ITC, SPIC, FACT, TATA Chemicals Ltd., Stanes etc.) are mainly involved in private extension relating to their products, namely seeds, fertilizers, pesticides, machineries, processed food etc. The Mass Media is the most powerful tool in agricultural extension to cover a large number of people at a time with less cost. At present there are about 45000 newspapers and periodicals in the country and the total circulation (copies) of the Indian Press is about 12 crores. There are more than 500 agricultural magazines. The radio coverage of the country is 98 per cent of population and 90 per cent of geographical area whereas the TV coverage is 87 per cent of population.

Recently information technology is also employed in private extension services. Typical examples are the information kiosks of Infovllages in Pondicherry established by the MSSRF, information kiosks (Choupal) of ITC, kiosks run by Samaikya (a private extension provider of Andhra Pradesh) etc. Also there are numerous agricultural websites (e.g. ikissan.com, krisvihar.com, agriwatch.com, commodityindia.com, plantersnet.com, aquachoupal.com etc.).

Services Rendered by Private Extension Agencies

1. Information

Prices of commodities, details of manufacturers, buyers, nurseries, input dealers, stock position etc., credit, sources and formalities; cultivation of crops and package of practices; solution to specific problems; tele-conferencing with experts; e-commerce; health education; video film production; publishing agricultural magazines; organizing agricultural exhibitions; demonstrations; training; film shows; information on government benefits etc.

2. Consultancy

Project planning and implementation for individual farmer or groups or organizations.

3. Input Supply

Seeds, seedlings, fertilizers, pesticides, weedicides, credit, labour etc.

4. Infrastructure

Transport, cold storage, storehouses, machineries (rent and repair), processing, packing etc.

5. Technical Service

Soil testing, diagnosis of diseases and pests, leaf analysis, water analysis, fertilizer testing etc.

6. Marketing Services

Buying and selling agents.

7. Other Services

Crop, livestock and life insurance agents.

8. Enterprises

Contract farming, seed production, procurement, packing and marketing, organic farming etc.

Streamlining and Strengthening of Private Agricultural Extension

1. Promotion of Private Extension

The government should support private extension service provider till break-even point or else interest free credit. Government should encourage group or cooperative activities. Concessions in sales tax, income tax etc. to encourage extension services may be considered. Local infrastructure such as sanitation, electricity etc. may be provided locally by the government.

2. Regulation of Private Extension

Private extension activity has to be promoted linking the employment for unemployed graduates, post-graduates and retired professionals for registration and financial back up. Professional graduation may be made mandatory. Extension techniques duly approved by competent department only be allowed. Only reasonable rates must be charged from the farmers. Agricultural extension system must conserve the natural resources and preserve ecological balance. Both central and state government should formulate policy guidelines specifying the role of private agencies in agriculture through appropriate legislation.

3. Innovation in Private Extension

Private extension service provider may be given access to public research information and technologies developed by government agencies. Each agricultural business company can be specialized in one extension activity. Multidisciplinary approach may be encouraged. Farmers' super market may be promoted. Agri-business companies may popularize contract farming. There should be a forum where innovation in private extension may be documented and shared among agri-business companies. Tele-consultancy may be promoted. Public information may be shared with private extension agencies. Companies can adopt villages to provide extension services. Involvement of women and marginal farmers in private extension is to be encouraged.

Why Privatization in Agricultural Extension

Looking at the population growth, the policy-makers are planning to double the food production by next ten years. Keeping this in view, the public extension system will not be able to meet the requirements of the farmers on *economic and efficiency grounds*. The public agricultural extension services are funded by the state and are widely viewed as supply-driven rather than demand-driven. Privatization of agricultural extension seems to be a solution to achieve the target of doubling agricultural production. Private extension is also necessary to provide self-employment to a large number of graduates passing out of the agricultural universities.

Criticism Against Privatization of Extension

The major drawback pointed out in private extension is the lack of human face and exploitation of marginal farmers. The private extension services usually concentrate in high value crops, namely horticulture, commercial crops etc. which have both domestic and global markets. Only the so-called progressive and wealthy farmers will be utilizing these services as they can afford to pay for the services. Small and marginal farmers who take up production of cereals and pulse crops are likely to be totally ignored. This will create a great imbalance among the farmers and the gap between the rich and poor farmers will be widened.

Even though privatization of extension may be successful in European and western countries, wherein only 5 to 10 per cent of their population are professing farming in contrast to 70 per cent of our population dependent on agriculture, it effects under Indian scenario may be different where small and marginal farmers practicing subsistence agriculture dominate. Under this context, privatization of extension will make additional financial burden to the farming community of the country, which is already financially weak.

Privatization is bound to neglect these less profitable crops urging the government to rely on imports or external markets for our food requirement.

There is also apprehension that privatization may not care for sustainability, instead may advocate exploitation of natural resources to the maximum extent. It is likely that the farmers may become a victim of unsustainable development by the activities of the poorly competent private extension agencies.

Adjustment with Present Public Agricultural Extension

The public agricultural extension system now lacks the support of market research, value addition to the produce, institutional reforms, professionalism and capacity building. If the public extension system is properly strengthened, it could serve the farming community better and build the confidence of the farmers which will serve as a safeguard against the over exploitation and evil effects of private extension agencies. The co-existence of public and private extension systems with sound linkage will be ideal for a developing country like India where subsistence agriculture dominates over commercial agriculture.

II. CYBER EXTENSION

Cyber extension refers to the use of Information Technology (IT) in the field of agriculture especially in transfer of technology.

Information technology has connected the world globally and is changing our life style and social consciousness dynamically. Of late, it has emerged as the best tool for information sharing and mutual communication. None of the walks of life has been left untouched by the IT sector, be it grain threshing or global business. Agriculture has also been greatly influenced by IT in the present era although the share of IT in agriculture is only 1.3 per cent.

The farmer, towards whose welfare a huge government

machinery and industry is devoted, still suffers from the absence of right information at the required time. The farmer requires timely information on weather forecasts, sowing time, availability and recommendations of inputs, availability of credit, expert advice on maintaining his crop in healthy condition, information on markets and on all other areas of interests to him and his family. Despite the best efforts and expenditure, the conventional government machinery could not meet these requirements satisfactorily. Herein lies the potential of IT, which can efficiently address the concerns of farmers stationed at even remote locations.

Institutional Support for Farm IT

It is to be noted that the farmer is in no position to use IT directly. The literacy levels, language barrier, cost of computers and poor infrastructure make it impossible for individual farmers particularly small farmers to directly adopt IT. This calls for institutional effort to establish necessary IT based services to farmers.

The Ministry of Agriculture, Government of India has set up Kisan Call Centres, which shall serve as easy knowledge access point for farmers. This programme was inaugurated by the Prime Minister of India on 21 January 2004. The farmers can call up these call centres on toll free telephone number and enquire about the various queries/problems related crops, seeds, fertilizers, agricultural commodity prices, pesticides, horticulture, veterinary and animal sciences, fisheries, forestry etc. These centres (Level I) are supported by Level-II Subject Matter Specialists available at various research and extension establishments, connected through high bandwidth telecom and computer connectivity. The Government of Kerala is operating a similar project entitled "Kisan Information Systems, Services and Networking, Kerala" (KISSAN-KERALA).

IT Kiosks

Realizing the inability of farmers to access information, many governmental and non-governmental agencies have come forward to establish IT kiosks in rural areas in order to serve the farming community with the latest advancement in information technology. Each IT kiosk is provided with a PC and modem with telephone dial-up to connect the databases/agricultural portals. The farmer can take the assistance of the operator of the kiosk for nominal charge. Typical examples are the kiosks established by various state governments, infovillages of MSSRF, choupals of ITC etc.

Network Requirements

The major bottleneck in spreading e-culture to rural areas is related to connectivity, i.e. difficulty in providing access points to get connected to the databases which are in selected locations. Since dial-up lines are very slow, other viable options are required to be explored. Unfortunately, the alternatives are expensive and may not be feasible. One silver lining is that several private operators are connecting important cities with fibre optic, which provides a very reliable and fast access. Since these will pass through rural India, it is feasible to explore the possibility of tapping this potential by laying the last mile connectivity. Once it is done, substantial segment of rural India can access the IT based services discussed above.

Community Owned Rural Internet Kiosks

The concept of community owned rural Internet kiosks is being operationalised in many states of the country. Since these kiosks are connected to a state network, they can serve better with authentic database available with government agencies. Some of the services rendered by community kiosks are :

1. *Agriculture produce market rates*: Prevailing rates of prominent crops at the local and other recognized markets/ auction centres around the country are available on-line for a nominal charge.
2. *Copies of land records*: Documents related to land records are provided on the spot at a nominal charge.
3. *On-line registration of applications*: Villagers had to make several visits to the local revenue offices/krishi bhavans to

file applications for obtaining certificates/claims. This can be avoided with on-line facility.

4. *On-line public grievance redress*: A complaint can be filed and a reply received within 7 days for a nominal cost.
5. *Village auction site*: It provides auction facilities to farmers and villagers for land, agricultural machinery, equipment and other durable commodities. One can put one's commodity on sale for a nominal charge.
6. *Other services*: On-line matrimonial advertisements, information on government programme, a forum for school children to ask questions, consultation with the experts, e-mail facility etc.

Areas of Information Technology in Agriculture

1. Access to Global Information System/Network (Internet)

Agricultural Portals

Today information on any field is now available in the global information system referred as Internet. A portal refers to a catalogue of records of related information on a particular area organized systematically and made available to the users on Internet.

Agricultural portals may be of either multipurpose or subjects specialized. A large number of governmental and non-governmental agencies have developed agricultural portals for the use of farming community. Subject specialized portals ensures that the client no longer need to guess the required information on the specialized field but can retrieve authentic information from the portal.

An agricultural portal has three main broad areas, namely (i) resource management for the management of natural resources for sustainable agriculture, (ii) products and industries which include information on agricultural products and industries/ commodities and (iii) agri-business which includes business information specific to agricultural marketing. These major areas are further broken down to broad subjects, specific subjects, governmental information etc.

2. Expert Systems in Agriculture

An expert system is an intelligent computer programme that uses knowledge and inference procedures resolve problems that are difficult enough to require significant human expertise for their solution. This software comes under the category of decision support tools. The programme imitates an expert by involving a client in a problem-solving situation, often providing a recommendation in response to the client's request for help in making a decision.

In general, an expert system will include (i) a knowledge base of domain facts and associated heuristics, (ii) an inference procedure or control structure for utilizing the knowledge base and (iii) a natural language user interface.

The domain constitutes a body of information widely shared and publicly available within the domain. The heuristic knowledge is mostly privately and individually held. It includes rules of thumb, judgments and sometime experience based guesses that typically characterize human expert level decision-making. In addition to knowledge base the system includes an inference or procedure often referred as inference engine. The remaining component of the expert system permits bi-directional communication, which is known as input/output system.

The difference between an expert system and a conventional computer programme is that the expert system manipulates knowledge while conventional programme manipulate data.

Some of the expert systems developed for use in agriculture are:

- Grain Marketing Adviser
- COMAX (for crop management in cotton)
- GOSSYM (crop recommendations based on soil fertility, weather data, pest management information etc.)
- POMME (for pests and orchard management of apples)
- SOYEX (soybean oil extraction expert system)
- FINDS (farm level intelligent decision support system for optimal machinery management)

3. Remote Sensing and GIS

Remote sensing and geographical information system have been employed in agricultural management especially in the characterization of soil, vegetation, pest detection, water stress etc. However, the availability of fine spatial resolution, near real-time data has limited its application in the past. Now, satellite imagery is available to meet the resolution and temporal requirements for agricultural management and the techniques are being successfully employed in the field of precision farming especially in developed countries.

REFERENCES

MANAGE. (2000). *National Seminar on Private Extension: Approaches and Challenges in the Millennium – Abstracts* (ed. Chandrashekara, P.), National Institute for Agricultural Extension Management, Hyderabad, 70p.

MANAGE (2001). *Private Extension: Indian Experiences* (ed. Chandrashekara, P.), National Institute for Agricultural Extension Management, Hyderabad, 226p.

Chapter - 6

Privatisation of Agricultural Extension

Perspectives and Prospects

N.P. Kumari Sushama and *C. Bhaskaran*

Introduction

Privatizing extension, as one strategy for providing efficient services for farmers is becoming prevalent in developing countries including India. Many developed countries have gone far ahead with privatizing extension and the outcomes have been mixed. Extension privatization is under active consideration in India and many State Governments have initiated measures to achieve this end. Savas (1987) defined privatization as the act of reducing the role of government or increasing the role of private sector in an activity or in the ownership of assets.

According to Bloome (1993) private extension involves any person in the private sector who delivers advisory services in areas of agriculture and is seen as an alternative to public extension. Swanson and Clarr (1984) are of the view that "Extension needs tc teach farmers the management and decision-making skills, help rural people, develop leadership and organization skills, enabling them to organize better, participate in co-operative credit societies and other support organizations."

Private firms are involved in Extension to (1) promote sales of production inputs or services, (ii) assure continuous supplying and quality of agricultural products; and (iii) promote or project returns on investment in farm. In the Indian context, the crucial issues to be looked into are the ability of the farmers to pay for the extension services and how to generate the benefits of extension as 'private' and 'public' (Sulaiman and Gadewar, 1994).

Table 6.1: Types of organizations providing extension services in India

Sl. No.	*Types*	*Examples*
1.	Line Departments	Departments of Agriculture, Horticulture, Animal Husbandry etc. of various state governments
2.	Universities	Directorate of Extension and other colleges of State Agricultural Universities and Agricultural Schools of Open University (YCMOU, Nasik)
3.	Krishi Vigyan Kendras	Krishi Vigyan Kendras sponsored by the ICAR in various districts
4.	Farmers' Associations	Maharastra Grape Growers Association Kerala Mushroon Growers' Association
5.	Producers' Cooperatives	Milk Co-operatives in different parts of the country
6.	Research Institutions	Various research units of ICAR and SAUs mainly through their outreach programmes
7.	Input Industry	Seed companies such as Ankur Seeds, Nagpur; Kumar Gentech, Pune; Messina Beej Pvt. Ltd., Bihar; Fertilizer Companies such as IFFCO, KRIBHCO, FACT, Indo-Gulf etc., Pesticides and farm machinery firms
8.	Consultants	Individual consultants and consultancy firms such as Green Plus, Nasik
9.	Non-Governmental Organisation	NGOs working in the area of agriculture and rural development such as BAIF, Pune; PRADAN, Bihar' and GSSS, Udaipur etc.
10.	Commodity Boards	Rubber Board, Spices Board etc. under the Ministry of Commerce
11.	Marketing Boards	Maharastra State Agricultural Marketing Board
12.	Media Print	News papers (Agricultural pages of most language dailies)
13.	Media and Audio and Visual	All India Radio and Doordarshan through its farm programmes and satellite channels such as E-TV (Telugu) through their farm programmes
14.	Others	Autonomous agencies in specific areas (Command Area Development Agency) and crops (Kerala Horticultural Development Programme); Banks through their field officers; Panchayat Samities through their Agricultural wings; (DOA); Irrigation Management Training Centres /Water and Land Management Institutes under the Irrigation Departments; Agro-processing Companies such as Pepsi Foods (Punjab) and ITC (Andhra Pradesh for its contract growers; and Internet

As far as agricultural extension is concerned, privatization is used in the broadest sense of introducing or increasing private sector participation, which does not necessarily imply a transfer of designated State owned assets to the private sector. Private extension services are primarily are of two types. The first is the entirely private type which is directly involved in farming activities through consultants, agri-business, agricultural input industries etc. The second type consists of farmer's organization, NGO's etc. which remain largely dependent on government subsidies.

The important contributions made by agricultural extension in promoting agricultural development and increasing food production resulted in increased interest in extension during the last few years (Van den Ban and Hawkins, 1997). Extension services, mostly public funded worldwide until a decade back, are increasingly coming under private domain. The increasing inability of the governments to adequately fund its extension machinery is the real force behind the search for alternative approaches such as 'cost sharing' and 'privatization'. India's extension system is considered to be the largest in the world. Currently, five extension systems are prevalent in the country: First Line Extension System (ICAR/SAU), National Agricultural Extension Service/ T&V system, Special Extension Programme on specific crops, Rural Development programmes and Extension programmes of NGOs each having its own mandate.

Problems of Privatization in Agricultural Extension

1. Though the advice from a privatized system may be more effective because the farmer can select an adviser who is best able to help, privatization of extension service may hamper the free flow of information (Van den Ban and Hawkins, 1997).
2. The commercial interests of these private agencies jeopardize the efforts by public research and extension systems towards eco-friendly and sustainable agriculture.
3. The HRD role of organizing, motivating and guiding farmers for empowerment will be sidelined.
4. Contact between farmers and extension agents gets declined.

5. Inevitable fallout of such an arrangement will increase regional imbalance.
6. Small and medium farmers who do not have access to public extension services will be eventually displaced from commercial production (UNDP, 1991).

Prospects of Privatization in Agricultural Extension

1. As Indian extension is more of an information delivery system than an information acquisition system, the scope of privatization and cost recovery are very limited.
2. The public extension system must make earnest efforts to organize farmers' groups and motivate and direct them so that they can demand appropriate and better extension services on a cost sharing basis.
3. Another important possibility is empowerment of small and marginal farmers through cost effective intensive use of electronic communication technologies.

Comparison of Public and Private Extension Service

The debate concerning the role of public sector agricultural extension was initiated in the 1980s. It gathered momentum in the 1990s.

Problems Observed in Public Extension System

1. Public extension services are widely viewed as supply driven rather than demand driven.
2. Commercialization of agriculture gave rise to specialized client and demand for location specific extension services which are not catered by public extension system.
3. Public extension deals with a large area, large population and diverse cropping pattern.
4. High cost, low impact of extension programmes, growing conflicts between farmers interests and policy, poor motivation of staff and conflicting roles are observed in public extension.

5. Insufficient face to face contact between extension worker and farmer.
6. Inadequate funds for operational purpose.
7. Majority of the extension services is curative in nature.
8. Inadequate technical qualifications of VLW.
9. Incomplete extension services.
10. Inadequate internal organization structure.
11. Inefficiency of extension personnel.
12. Inappropriateness or irrelevance of extension content.
13. Dilution of impact.

Claims Regarding Private Extension

1. Private extension reduces the economic burden of governments as in Netherlands.
2. Increases the efficiency of extension services provided up to the satisfaction of farmers.
3. Economic interest of the extension agent increases his involvement in providing extension services as in share cropping.
4. Privatization increases the accountability of extension agent as in the Extension Contract System in China.
5. Privatization increases the competency of the research system as in the Extension Contract System in China.
6. It is possible to prevent the farmers being exploited by private extension agents by proper legislations as in the Management Committees of Denmark.
7. Money spent by private consultants to provide the same extension services provided by public system is less due to high overhead charges in the latter case.
8. Government can ensure quality extension services by technically competent extension agents through legislation as in the Certification in Germany.
9. Government can privatize extension while concentrating more on other important areas as done in Tunisia.
10. Government can have maximum control over private extension activities as in Chile.

In India, more than 70 per cent of the operational farm holdings are small and marginal (with less than two hectares). The ability of the farmers to pay for the extension services is very poor. Moreover, 70 per cent of the net sown area in India is rainfed subjected to external vagaries with poor resources and weak infrastructural support. But this situation is qualitatively different in irrigated areas and even at present especially in commercial crops, the big farmers are receiving advice and other extension services from private consultancy/TOT firms. Considering these diversities in the production systems and farmers in India, a very cautious approach is warranted with regard to privatization of extension services.

REFERENCES

Bloome, P. (1993). Privatization lessons for U.S extension from New Zealand and Tasmania. *J. of. Extn*. Spring.

Savas, E.S. (1987). *Privatization: the key to better Government*. Chatham House Publisher. Inc., New Jersy. pp. 3-10.

Sulaiman, V.R. and Gadewar, A.U. (1994). Privatization of extension services — Implications in the Indian context. *J. Rural Reconstruction*. 27 (2).

Swanson, B.E. and Clarr, J.B. (1984). The history and development of agricultural extension in Swanson, B.E. (Ed.) *Agricultural Extension—A Reference Manual*. FAO, Rome. pp. 1-19.

UNDP (1991). Agricultural Extension Programme Advisory Note. New York.

Van den Ban, A.W. and Hawkins, H.S. (1997). *Agricultural Extension* (2nd Ed.) John Wiley and Sons. Inc., New York. pp. 254-58.

Chapter - 7

Private Extension—The Urgent Need of Indian Agriculture

M.V. Sajeev and *B.K. Narayana Swamy*

Introduction

The Indian farmer is rapidly changing from subsistence to commercial lines. His dependence on Agriculture now has the challenge to meet the diversified needs growing out of modernization. Keeping on line with these changes, the expectations of farmers on extension have also changed. Even though India is self-reliant in food production, our population is not nutritionally secure. There is a need for more focus of extension on low cost technologies, high production technology, Value-addition through processing quality, price competitiveness, marketing and use of IT in Agriculture. The pressure of the changing roles and additional responsibility on public extension will add to its whole lot of existing woes like wide ratio between extension worker and farmer, inadequate infrastructure and finance. This existing gap is always enlarging. So it is in the interest of Indian farmer that private extension has to be encouraged. Keeping this in view, this chapter deals with the concept of private extension, why privatization came into force, its expected benefits, possible dangers, elements involved, expected roles and strategies for privatizing extension.

What is Private Extension?

Private extension is the services rendered in the area of agriculture and allied aspects by extension personnel or organizations, as supplementary or alternative to public extension services and service on payment basis.

Why Privatisation ?

The most important reasons behind the issue of privatization of extension system are explained below:

1. To Reduce Financial Burden on Government

The first and foremost reason behind privatisation can be attributed to the declining trend in government expenditure for extension in several countries over the last decades. Financial burden of governments has forced to make sharp reduction in the budget of public extension programmes.

2. Disappointing Performance of Public Extension Service

The disappointing performance of public extension service is due to aspects like :

1. Ineffective extension work.
2. Economically non-viable transfer of technologies.
3. Unmatching with farmers needs and little consideration for cost effectiveness.
4. Less competent extension worker with no accountability to the farmers.
5. Bureaucratic nature of Extension workers.
6. Target oriented programmes mostly imposed on farmers.
7. Inability to answer the entire demand of farming community.

3. Low Coverage of Public Extension System

This occurs in terms of time and services, i.e. inability of public extension system to reach all the farmers, all the time and inability to provide all the services needed by farmers.

4. Wide Extension Worker : Farmer Ratio in India

The extension worker : farmer ratio is very wide in India, i.e. 1 : 1000. This further widens due to reasons like :

1. At least, 25 per cent of extension workers are administrators/ supervisors who are not directly in touch with farmers.
2. With the remaining extension workers, at least 50 per cent of the time goes for administrative work, official correspondence, reports and travel to reach the villages.
3. Excluding the leave period, holidays, an extension worker attends office for about 250 days in a year.

5. Confined Role of Village Level Extension Worker

At present the role of VLEW is confined to providing advisory services to the process of production. But the fate of farmer is decided later in processing and marketing where VLEW has limited role. The VLEW's role is restricted to recommendation of technologies but whereabouts of inputs, information on price and infrastructure for storing, processing etc. is left to the farmer. The VLEW is responsible to provide information but not accountable to the consequences. Due to lack of accountability, the importance of extension worker is reduced in the eyes of the farmers.

6. Commercialisation of Agriculture

In the recent past, most of the developing countries including India agriculture is shifting from mere subsistence level to commercialised agribusiness. Above all, the challenges and opportunities of globalization and liberalization era demand effective alternative extension approach.

7. To Reduce Marketing Problems

At present farmers are facing lot of marketing problems. Hence the specialised nature of agriculture will demand quick and technically sound advice, based on scientific analysis with appropriate marketing information. Existing public extension system is not capable of meeting these challenges.

Expected Benefits from Private Extension System

1. Generates new, additional income.

2. Provides demand driven service.
3. Increases the voice and commitment of farmers.
4. Extension service becomes more cost effective with efficieny.
5. Complement or supplement the role of public extension system.
6. Extension personnel become more client accountable.
7. Increases staff professionalism.

Possible Dangers of Private Extension System

1. Private Extension Service may avoid small and marginal farmers.
2. Private extension may focus only on areas having favourable environment.
3. Private extension may become less educative and more of commercial oriented.
4. Private extension may focus less on foodgrain production and more on cash crops.
5. The contact between extension worker and farmer may be fully commercialised.

Expected Model of Private Extension System

Expected model of private extension system should cover :

1. *Objectives*
   (a) Profit maximisation
   (b) Increasing the efficiency
2. *Target group*
   (a) Only those who can pay
   (b) More committed
   (c) Highly careful
   (d) Actively participating
3. *Mandates*
   (a) Technologies
   - Location specific
   - Demand driven

- Timely
- Profitable

(b) Input supply

- Quality, timely and adequate.

4. *Organisation*
   (a) Private consultancy
   (b) Commercial agro-based firms
   (c) Input agencies
   (d) Farmers organisations etc.
5. *Extension professional*
   - Accountable to farmers
   - Highly professional
   - Highly motivated
6. *Funds*
   (a) Client's contribution
   (b) Development agencies
7. *Extension methods*
   (a) Advisory in nature through personal communication.
   (b) Participatory approach to generate new income.
   (c) Less use of group and mass contact.

Expected Roles of Private Extension System

1. Farm advisory services for profit maximisation of clients
2. Timely input supply for better production
3. Providing market information and market intelligence
4. Processing the client's produce
5. Marketing the client's produce
6. Providing credit facilities to the farmers
7. Providing infrastructure facilities like transport and storage

Possible Strategies for Privatizing Extension System

1. Commercialization of Extension Services

Here complex, demand driven technologies will be provided for a particular cost. These are the services which public extension system often fails to provide.

2. Introducing "Contract Extension System"

Public extension system can make contract with registered private agricultural consultancy agencies to transfer the agricultural technology.

3. Introducing Share Cropping System

Private extension agents are provided with remuneration in the form of share crop. It will increase the extension worker's accountability and commitment to the service.

4. Giving Partnership Rights and More Responsibility to Private Sector and NGOs

Private sector and NGOs are entering in a big way in recent years to provide agricultural consultancy. They may be given more responsibility in agricultural technology transfer.

5. Selective Withdrawal of Public Extension Service

Selective withdrawal should be done area-wise and commodity-wise. Areas having favourable environment like high soil fertility, high irrigation potential, satisfactory infrastructure facilities and commercial crop growing areas. Responsibility regarding commodities, which give high profit to farmers, can be given to private sector.

6. Creating and Strengthening Farmers Groups and Cooperatives

Through farmer groups, co-operatives, extension agents are appointed and members will share cost. For this purpose existing village co-operative, clubs, mahila mandals and water management committees can be used.

7. More Emphasis on HRD

To increase the professional competency, more emphasis should be given to develop ability and willingness among extension professionals.

Conclusion

The reasons explained for the emergence of private extension like financial constraint of government, disappointing performance of public extension system etc. will only become more severe in the coming years. Considering the positives and negatives of private extension a strategic decision has to be taken and implemented in the shortest possible time. Private extension will make extension services more decentralized, farmer participatory, flexible and provides ownership rights to the clients. The coexistence of public and private extension service is surely going to help the ultimate consumer, the farmer, through competitive, cost effective and efficient service.

REFERENCES

Ameur, C., (1994), Agricultural Extension: A step beyond the next step. *World Bank Technical Paper*, No: **247**, The World Bank, Washington DC., USA.

Anonymous, (2000), Proceedings of the *National Seminar on Private Extension : Approaches and Challenges in the Millenium.* July 28-29, 2000. MANAGE, Hyderabad.

Antholt, C.H., (1994), Getting ready for the twenty-first century – technical change and institutional modernization in agriculture. *World Bank Technical Paper*, No:**217**, The World Bank, Washington DC., USA.

Dinar, A. (1996), Extension Commercialisation: How much to charge for extension services. *American Journal of Agricultural Economics,* **78** (1): 1-12.

Howell, J., (1985), *Recurrent Costs and Agricultural Development.* Overseas Development Institute, London.

Narayana Swamy, B.K., (2001), Human Resource Development among professionals as an approach and challenge for privatization of extension in India In Proceedings of the *National Seminar on Private Extension : Approaches and Challenges in the Millennium,* July 28-29, 2000. MANAGE, Hyderabad.

Rivera, W.M. and Gustafson, D.J., (1991), (eds.) Agricultural Extension: World-wide institutional evaluation and forces for change. Amsterdam : Elsevier.

Saravanan, R. (1999), A study on privatisation of agricultural extension service. M.Sc. (Ag. Extension) thesis (unpublished), Department of Agricultural Extension, UAS, Bangalore.

Senthil Kumar, N. Nomesh Kumar and Narayana Swamy, B.K., (1997), Privatization of extension as a viable option for globalisation of Indian Agriculture, Proceedings of international symposium on globalising Indian Agriculture : Policies and Strategies for 21st century, December : 27-28, 1997, Chennai, Tamil Nadu.

Chapter - 8

Attitude of Farmers Towards Public and Private Extension for Sustainable Production to Meet the Challenges of WTO

S.K. Meti, Ravi M. Sambranith, J. Tulasiram and *K.N. Doddamani*

Introduction

Indian farming is in a transition stage changing from substance farming to surplus and export oriented farming as it is one of the member countries of World Trade Organization (WTO). For this purpose the production of different agricultural commodities has to be increased in terms of quality and quantity at reduced cost of production to compete globally. Public research and extension played a vital role in bringing about the green revolution. In the post liberalization era however, the extension focus important challenges in the area of capacity building, relevance and accountability.

Capacity Building : Means of import sufficient knowledge and develop adequate skills among the extension personnel and subsequently in farmers, to produce high quality agricultural and allied products in large quantity so as to enable the farmers to get good price at local, national and international market, for raising their income. The dedicated extension work with positive attitude, full knowledge and technical know-how helps to energize the farming community for motivation and competition. Under the WTO regime, the reformation in public research, extension approach, a role performance of private actors like NGOs, voluntary organization, farming groups etc. have to be strengthened with technical knowledge, skill and modified their attitude to work parallel to the public extension.

Already we have the infrastructure, manpower, material power, technologies, money power. For integration of all these powers only the requirement is the management power, which is to be enhanced in extension personnel and farmers for judicious use of natural available resources for sustainable production. Unscientific method of using and transfer of technology leads to the soil degradation, low quality production, which are not in a position for acceptable in the national/international market. Further, environmental pollution and health hazards lead to spoils the human competitive environment.

Structural and functional change in agricultural extension is important to meet the needs and challenges of the new millennium. The globalization and liberalization era has made people to rethink the role of public extension and private extension for modification of old strategies in developing countries.

Human Resource Development

The extension professionals having background of extension system, i.e. public extension system for which dedicated personnel with managerial skills acts as critical input for successful agribusiness. Agribusiness managers should be familiar with business because ethics, socio-economic conditions too besides being proficient in agricultural knowledge and managerial skills.

The challenge today is to get the concept of agribusiness adopted in farm sector in general and by small and marginal farmers in particular, so that their share in the success is also ensured. The opportunities in agribusiness are enormous and can be easily encashed with indigenously available technologies. Now is the time for young farmers and agribusiness managers to accept challenge for mutual benefit. The attitude of small and marginal farmers towards improved agriculture enables them to acquire knowledge, skills for adoption and earn the profit. Therefore, the study was designed to know the information about attitude of small and marginal farmers with respect to improved agricultural technologies by using the following objectives.

Objectives

(1) To study the level of attitudes towards improved agricultural technologies by small and marginal farmers.
(2) To study the significant mean differences of small and marginal farmers between knowledge, adoption and attitude with respect to cotton, paddy and groundnut.

Methodology

The study was conducted during 1998 in TBP area of Raichur District of Karnataka State involving 120 small and 120 marginal farmers from 12 selected villages. The improved agricultural technologies like Land Management Practices, Crop Cultivation Aspects, Integrated Pest Management, Nutrient Management practices and subsidiary acceptation were considered for the study. Small and marginal farmers of TBP area gathered on the information attend towards improved Agricultural technologies. The independent variables for the study were age, education, experience, social participation, mass-media participation, social contact, innovativeness and risk orientation. The attitudes of small and marginal farmers were obtained by following the procedure of Likert (1932) and Edward (1969) The gathered information was analyzed and presented below.

Result and Discussion

The data indicate that, majority of the small and marginal farmers (56.67%) belong to favourable attitude towards improved agricultural technologies. However, the percentage of small and marginal farmers who had less favourable attitude was 23.75 per cent, which is a fact to be considered. A less number of small and marginal farmers (19.58%) belonged to high attitude category, which is not encouraging.

Further, it was observed that, small farmers mean attitude score was more than the mean attitude score of marginal farmers. The mean attitude score difference between these categories was 6.13 per cent, which is significant at 5 per cent level. Hence, the null hypothesis that there are no difference in the level of attitude towards improved technologies was rejected.

It could be inferred that, it is necessary to change the attitude of small and marginal farmers towards improved agricultural technologies by greater involvement of these categories in extension activities. Specially designed for small and marginal farmers.

Significant mean difference between knowledge, adoption and attitude score among the small and marginal farmers with respect to groundnut, cotton and paddy

It is found from the data that there was no significant difference in the knowledge level between small and marginal farmers with respect to groundnut, cotton and paddy cultivation practices. This might be due to almost similar socio-economic conditions except land holding, the social status also remains same in the society and also both get subsidy from the Government. This finding was contradictory to the results obtained by Ramachandran (1974) who found that there were significant differences between small and marginal farmers in their knowledge about high yielding varieties of groundnut, cotton and paddy.

In case of adoption level of groundnut, cotton and paddy crops practices, the significant difference in the mean adoption score was observed between small and marginal farmers. This might be the fact that, the bigger land size of small farmers might have influenced them to adopt technologies and increase yield level. In other words the economic status of small farmers will encourage them to adopt new ideas that contribute to their income. Hence the results.

Further, it was observed from the data that, there was significant difference between the small and marginal farmers with respect to attitude towards the improved agricultural technologies. This might be due to the fact that, small farmers are having opportunity to increase their yield level due to more land as compared to marginal farmers. Usually small farmers were having more favourable attitude towards improved agricultural practices than the marginal farmers.

Role of Public Extension

- To develop practically efficient and effective public extension system, i.e. demand driven, well integrated with research, and financial and motivational aspects.
- To utilize the expertness of NGOs to help all types of farmers especially the resource poor to get organized into groups which will help to them access new technologies and to strengthen their voice in providing feedback to research and extension.
- To sharpen the focus of the public on agriculture technology generation system with an emphasis on location specific, system based and sustainable technologies by encouraging participation of the private sector, i.e. Quality seed and input supply capacity to expand their role in transfer of technologies to the farmers.

Weak Linkages with Public Extensions

The entire extension approach to the rural and agricultural development was inconsistent, frequently changing and replace with uncertainty and confusion. Even after 50 years, we could not give definite direction, develop right concepts, provide permanent infrastructure and ensure minimum basic physical facilities.

- The research centre are not still fully equipped in relation to technology development.
- The extension workers could not visit villages frequently and establish direct rapport with the farmers.
- The extension workers could not update their knowledge and develop professionalism to perform the major role of communication and achieve high level of competence and credibility.
- In rapid development, several private institutions and agencies were created to support and strengthen the extension system for the common cause of helping the farmers.

In view of the weak links, the involvement of NGOs were

started working with specific objective in specific area. The NGOs level of credibility is high due to sincerity and honesty of purpose and dedication to the cause of rural development.

The Major Constraints in Public Extension that Need to be Addressed by ITD

- Multiplicity of technology transfer system without integration each other.
- Narrow focus of the agricultural extension system to the public.
- Lack of farmers focus, weak feedback and inadequate technical capacity within the extension system.
- Need for intensity by the farmers on need based training.
- Weak research, extension, farmer and market linkage is missing.
- Poor communication capacity.
- Inadequate operating resources and weak financial support.

The current institutional framework for technology transfer with its institutional and operational constraints severely limits the effectiveness of the public extension system to disseminate location specific technology to different socio-economic groups of farmers. Also because of policy changes in the wake of the economic liberalization. The private sector and commercial farmers are taking more responsibility for technology dissemination and technology adoption without proper future thinking. This dynamic situation calls for systematic reforms of the agricultural technology system (ATS) both to strengthen linkages between research, extension and farmers and also to achieve better system integration among the agencies involved in technology transfer. The generation of such demand driven technology is possible only if the research agenda is drawn on the basis of real life field problems. One way to ensure this would be to follow a participatory research approach where researchers, extension agents and farmers have a chance of expressing their observation and concerns to each other. Otherwise, the luxury of purely academic research will remain a burden on the scant budget resources of the national and state governments.

Concept of Privatization

Bloome (1993) indicated that, private extension involves personnel in the private sector, who delivers advisory services in the area of agriculture and is seen as an alternative public extension. Whereas, Van den Ban and Hawking (1996) state that, farmers are expected to share the responsibility for this service and pay all or part of the cost.

Saravanan and Shivalingegouda (1999) operationalised in the following manner, privatization of agricultural extension service refers to the services rendered in the area of agriculture and allied aspects by extension personnel working in private agencies or organizations for which farmers are expected to pay a fee and it can be viewed as supplementary or alternative to public extension service.

Role of Private Extension : Para-Extension Professionals

- Para-extension worker like, private company, NGOs, voluntary organization and farmer groups, acts as supplementary/substituting public extension.
- Public funds to support para-extension workers through training and capacity building, help in procuring loans from credit institutions, linkages with SMSs in line departments and SAUs.
- Public funds to support para-workers (Mitra Kisan and Mahila Mitra Kisan) through payment of honorarium.

Private Extension

Agri-clinics: Supporting an employee who is agricultural graduate to set up commercially viable agri-business through suitable training and loans to become service providers to the farming community.

Input Dealers: Through their associations to under-go periodic training and orientation programme to be able to render technical correct advice to farmers.

Corporate Sector: Provide enabling environment by reduction of constraints and simplification of procedures to enable contract farming wherever possible.

Co-financing Extension through Public Private Partnership

Competitive Extension Grant Fund: So which NGOs, Farmers organization foundation etc. can apply for undertaking specific extension tasks with the help of public extension personnel.

Information Technology and Agricultural Extension

In the world of globalization, an information technology (IT) plays an important role for sweeping changes in socio-economic development. At present, benefits of this technology is available primarily to urban elite masses. Rural communities still and will continue to have difficulty in accessing crucial information due to their poor understandable capacity, so as to make timely decisions. There is a gap between the information rich and information poor community. This gap has to be considered as important by public extension in the WTO regime.

However, application of such information technology (IT) is constrained by lack of complementary inputs (equipment, power, etc.), appropriate organizational and institutional infrastructure, information management, skilled human resources, and opportunities for technology transfer etc.

- It is a theoretical extension approach not productive and not physical one.
- It is only limiting to literate and elite farmers useful communication media.
- Still majority of extension personnel are not advanced in information technology operation and timely use of information.

Capacity Building of Farmers by Public and Private Extension System

- Promotion of direct interface between all categories of farmers and scientists.

- Use of more participatory approaches: This would increase the capacity of the farmer's to think and rethink analytically and find solutions for the problems on their own.
- Acquisition of skills by farmers.
- Assessing of farmers needs and skills for further development.
- Distinguish different dimensions of training such as awareness, knowledge, skills for reinforcement of small and marginal farmers by using appropriate communication channels and methods for capacity building of farmers.
- Different kinds of technologies and advice required by different categories of male and female farmers with transfer of technology mechanism (e.g. mass-media, and group) they prefer during different phases of awareness, trial and adoption of new skills and technologies.
- Use of information technologies for improving the quality and accelerating the transfer and exchange of information.
- Organizing training programmes on system based and sustainable technologies such as IPM and INM.
- Capacity building and skill upgradation through training of farmers would be largely in the farmers fields by scientists and extension personnel participating.
- Programme implementations through their institutions/user groups.

How to meet the WTO Challenges

- Small and marginal farmers empowerment with suitable technical knowledge and technical skill for cultivation of improved crops. This is possible only with effective and meaningful combination of different extension teaching methods.
- Extension worker should be strong for imparting practical useful skill through effective and timely communications.
- Farmers should be outside field exposure to know and understand the various techniques and field operations carried out by other farmers.

Future Area of Development

- Reorient the WTO standard to the farmers socio-economic level.
- Reorient the extension system to present status of WTO.
- Our communication must be productive and market oriented.
- Strengthen the market information sources including price, storage and transportation.
- Link with weather forecast information.
- Concentrate on quality and value added productions.
- Post-harvest technology should be disseminated.
- Educate the farmers about grading and other market procedure.

The Role of Third Extension Generation

- It says that farmers already possess much knowledge but they possess traditional ways of solving problems.
- Knowledge cannot be transferred; One cannot learn some one else's knowledge, one can only create one's own knowledge.
- Farmers are already solving their own problems without the aid of the extension worker. They are not resource poor in terms of learning.
- Create a knowledge—to question, to analyze, to test, possible solution for themselves.
- The interactive approach may be slower than directive and responsive approaches. It enables the farmers to become a independence of the extension worker.

Improved communication and information access is directly related to social and economic development. However, it is observed that, the rural population still have difficulty in accessing crucial information in forms they can understand in order to make timely decisions. There is a concern that, the gap between the information rich and information poor is getting wider.

According to the IT Task Force recommendations "The Government shall take necessary steps to boost IT for agriculture and integrated rural development".

By 2008, any farmer in remote village can on demand get the following information :

1. Cropping strategy for farmers field based on soil, weather, and fertilizer and pest management models.
2. How and where to get proper seeds or nursery plants?
3. Prevailing prices of various farm equipments and products.

One would like in an ideal situation the following to happen :

1. An agriculture information centre in each village.
2. Interactive exchange of information for planning and day-to-day operation by the farmers.
3. Availability of all the extension services on demand.

It is essential that information availability is demand driven rather than supply driven. The challenge is not only to improve the accessibility of communication technology to the rural population but also to improve its relevance to local development. The on-going IT revolution has opened huge opportunities for providing access to information as well as to interactive distance learning in rural India.

REFERENCES

Edwards. A.L., (1957), *Techniques of attitude scale construction.* Appletion Century Crofts, Inc. New York.

Edwards. A.L. (1969). *Techiques of attitude scale construction.* Vikiks. Feffer and Simons Pvt. Ltd. Bombay, India.

Likert, R., (1932). A technique for the measurement of attitudes. Archives of Psychology, No.140.

Ramachandran, P. K., (1974) .A multivariate study of information source by medium and small farmers. Ph.D. (Agri.) Thesis, (unpublished), Division of Agriculture Extension. IARI, New Delhi.

Siddaramaiah, B.S. and Jalihal, K.A., (1983), A scale of measuring extension participation of farmers. *Indian Journal of Extension Education*, 19:74-77.

Section - II

Extension Approaches

Chapter - 9

Self Help Groups—An Effective Extension Strategy for Rural Development

B. Narayanaswamy, K. Narayana Gowda, M.S. Nataraju and *M. Manjunatha*

Introduction

The agricultural development in India has undergone many ups and downs in the last 50 years. Many planned programmes introduced from 1952 onwards brought credible contributions for increased agricultural productions and we were able to achieve self-sufficiency in food production including buffer stock to meet many emergencies. But the living conditions of large section of farmers did not improve correspondingly. The cost of production has increased steeply but not the net income although productivity has increased several folds. There are many factors that have contributed for this dismal situation and most important ones are :

1. Increase in cost of inputs including labour wages;
2. Prices of outputs did not increase matching with production costs;
3. Many production assets are not affordable by individual farmers like pair of bullock, improved implements, plant protection equipment, threshers etc.;
4. Influence of World Trade Organization (WTO);
5. Lack of timely and adequate information on technological advancement;
6. Inadequate credit support for many low resource base farmers;

7. Uneconomical land holding size and indiscriminate fragmentation;
8. Less empowerment of farm women and youth.

All the above factors either individually or together/ combination affected the socio-economic conditions of farmers in India. The situation may go bad to work if proper mechanism is not evolved to address the aforesaid situation. More specifically, the present day family cannot be addressed individually. Therefore group approach is inevitable to bring about sustained development of farmers in India. Therefore an effort to be made to review the various steps undertaken by various organizations.

Empowering the social gamut in terms of participation or involvement in all development process is a vital aspect today. This helps people gain control of their lives through raising awareness, taking action and working in order to exercise greater control. Moreover, empowerment facilitates change and enables an individual to do what one wants to do. It inturn the feeling, that activates the psychological energy to accomplish one's goals. So any agricultural development programme should give more importance to help the farming community to empower themselves. They must be helped to help themselves through education and training for achieving sustainable rural development. There are several factors which are not favourable to rural community to emerge as entrepreneurs like lack of skill, knowledge, education and information, lack of self-confidence, lack of need for achievements, lack of inclination to take risk, lack of independent outlook which are essential for enterpreneurship. So to reduce social and technological risks faced by farming community, empowering them on group basis is the next best alternative. So far the limitations in implementations. At certain extent some NGOs grabbed this opportunities for entrepreneurship development on group basis and some organized Self Help Groups (SHGs).

What is SHG ?

It is a voluntary association managed democratically without any political affiliation (Ramesh, 1996).

Self-Help Groups are self-managed groups of poor men and women which have come into existence to mobilize financial resources through their savings and lend the same amongst themselves to meet the credit needs of their members.

Self-help group is not a static institution. It grows on resources and management skills of its members and their increasing confidence to get involved in issues and programmes that require their involvement in the public and private sectors. The members are automatically empowered economically, politically and socially in a holistic way for developing their own institution for common benefit through the SHGs.

These have been recognized today as the pillar stone of the poverty alleviation programme of this century. This recognition is reflected in the target of one million SHGs to be supported by re-finance from NABARD (Finance Minister's Budget speech 2000-2001) and Swarna Jayanthi Swarozgar Yojana (SGSY), the prime poverty lending programme which has replaced the Integrated Rural Development Programme from the year 1999-2000.

Why Aspects of SHGs

- SHG is a via-media for development of savings habit among the poor,
- It is an access to larger quantum of resources,
- It is a window for better technology/skill upgradation,
- Availability of emergent consumptional production credit at the door steps,
- Access to various promotional assistance,
- SHG is an assurance of freedom, quality, self-reliance and employment.

Some Aspects of SHGs

There are several success stories such as MYRADA, MSSRF, NABARD, OUTREACH, IRDAS, BAIF, DHAN FOUNDATION etc.

Different activities undertaken by rural people are studied and discussed hereunder :

1. Community Banking Through MSSRF

SHG is an empowerment process in which a community gains confidence and initiates economic activities. Rural poor can be brought under SHGs and their need can be met by micro-crediting and with other facilities. For example, one such activity to name is, 'community banking' through M.S. Swaminathan Research Foundation (MSSRF). Here micro-credit is given to a small and homogenous group of men/women having a similar socio-economic background is SHGs.

Through SHGs the MSSRF have taken up micro credit system in the activities like, Bio-pesticides, Mushroom cultivation, Boards/paper from banana wastes, Hybrid seed production, Fodder cultivation, Calf rearing, Goat rearing, Dairies, Poultry, Dal mill, Handicrafts, Vegetable cultivation etc.

All these activities are taken up within the framework of entrepreneurship, i.e., the technical, managerial capacities of villagers to organise themselves and produce goods for sustainable development. Micro enterprises start with mobilization of villagers, followed by group formation, capacity building and monitoring of the activities of the group.

2. Water-users Association of IRDAS

Of late the growing concern for the social science approach in all spheres of development projects has encompassed the need for people's participation. One such example is National Water Policy, 1987 emphasizes participation of farmers at various aspects and assistance from voluntary agencies for effective management. Karnataka State has taken steps to conduct Action Research Programme through NGO, i.e. ISARD (Institute for Studies on Agriculture and Rural Development, Dharwad) in TBP area. It has made a commendable success in making the farmers to formally organize themselves into a registered society to take over the responsibility of managing the caned irrigation.

Institute of Resource Development and Social Management (IRDAS) (1988) reported that water-users association in Sri Rama

Sagar Project (AP) succeeded by following the strategies like distinctiveness of the approach, direct and persistent communication with farmers, walking through village level meetings and individual and group contacts.

3. *Capacity Building Through SHGs by OUTREACH*

OUTREACH an NGO organizes community based programme, has made an impact on development in several areas, resulted in evolving capacity building institutions, multifunctional systems like Self Help Groups (SHGs) and secondary and tertiary community based organizations like the Cluster Level Associations (CLA) and federations. In 2001-02, 49 new SHGs are performing activities such as bulk purchase of agricultural inputs, micro enterprise development and marketing and in resource mobilization.

Name of the Project	*No. of SHGs*	*Total Members*	*Total Savings*	*Total Loan Amount*
OUTREACH Projects	1117	16954	1,23,55,846	4,20,64,028
Other NGO Partners	119	1537	13,76,070	1,38,80,795
Total	*1236*	*18491*	*1,37,31,916*	*5,59,44,823*

Here the advantages of community/group approach experienced are (i) Need-based credit, (ii) timely and effective support, (iii) low transaction cost, (iv) user controlled revolving fund, (iv) high repayment, and (v) credit accompanies by appropriate services,

4. *MYRADA : An Instrumental Agency in Capacity-building of Rural Community through SHGs*

Mysore Resettlement and Development Agency (MYRADA) is an NGO managing Rural Development Programme in 3 states of Southern India was started in the year 1968 in Mundagod of Karnataka for the rehabilitation of Tibetan refugies. There are 12 critical areas, which form part of MYRADA's programme in the field, in which MYRADA is making a contribution to development of theory, policy and practice. In some of these areas, notably in the SHG and watershed management strategies, MYRADA's pilot initiatives have influenced changes in national policies.

Since 1984-85, MYRADA has fastened SHGs of the rural poor. They form the power base of the poor which helps them to overcome obstacles in their paths to the river. In December 2002, there were 6886 SAGs (Self Help Affinity Groups) in MYRADA's projects, with 10,754 members, managing a total common fund of Rs. 555 million of which Rs. 229 million is savings and Rs.114 million interest earned on lending. NABARD supported MYRADA with a grant of Rs. 1 million in 1987 from its R&D fund, for the pilot initiative of looking SAGs to Banks. MYRADA's SHG strategy was officially accepted by Government of India in 1999.

5. *Bharathiya Agro-Industries Foundation (BAIF)*

Since 1967 BAIF is focusing on Socio-economically backward families through its multidisciplinary programmes such as ensuring people's participation through SHGs and community organizations and emphasizing on micro-level enterprises. This NGO works closer with people in a flexible and informal manner. The multi-dimensional approach of this NGO has formalized development activities like poverty alleviation, sustainable development, employment generation, conscientization of deprived classes and population control to make the people to prosper.

6. *DHAN Foundation*

DHAN foundation's community banking programme is based on the concept of SHGs. At the end of March 2002 the programme has promoted 7861 SHGs, with a membership of 1,22,118 families. The saving mobilized by these SHGs is nearly Rs. 235 million and loans given out till March is Rs. 430 million. The SHGs among them have generated a common fund of Rs. 44 million. Each SHG called Kalanjium.

Need of the Hour

We can ensure actual people's participation in SHGs. So it is a dynamic process where all the members of SHGs contribute to the attainment of group objectives, share the benefits from group activities, exchange information and experiences of common interest and follow the rules, regulations and other decisions made by the group. Though there have been many programmes to alleviate poverty, SHGs have done well in parts of the country and so it is necessary that more and more SHGs are encouraged to take up. Self help has emerged as a new paradigm for combating poverty and rural development. Hence all those concerned with Transfer of Technology can use these types of group approaches to reap the fruits of agricultural development.

REFERENCES

Anonymous, (2002), BAIF : In www.ngoindia.com

Anonymous, (2002), Annual Report, 2001-02, RRC, DHAN, Foundation, Madurai.

Banki, E., (1981), *Dictionary of administration and management,* Los Angeles, System Research Institute.

Fernandez, A.P., (1994), The MYRADA experience : Alternate management systems for savings and credit of the rural poor. MYRADA-Bangalore.

Katar Singh, (1991), *Managing common property resources.* Institute of Rural Development, Hyderabad.

MSSRF's Information Village. In : Poverty. Worldbank.org.

M.S. Swaminathan Research Foundation. In : www.mssrf.org.

Chapter - 10

Constraints in Adoption of Micro Credit Among Farmers

Priya R. Devi, C. Bhaskaran, N.P. Kumari Sushamma and *Elsamma Job*

Introduction

Micro credit is an informal arrangement for credit supply to the poor through Self Help Groups (SHGs), which is fast emerging as a promising tool for promoting income generating enterprises. Apart from NABARD, other national level funding agencies are also providing institutional arrangements for micro credit supply to ensure credit at lower rates for its beneficiaries.

Thus, shaping of rural credit system to supplement the small savings of rural population to account for cost of risk and uncertainty of farming is undertaken. As micro credit is widely appreciated, now focus must be on providing best services and introduction of innovative features. Recent reports highlights the importance for organising marketing efforts, technology upgradation and diversification of activities for groups associated with this programme (Nagayya, 2000). In this backdrop, this study was carried out with the objectives for identifying the constraints encountered in micro credit servicing and repayment.

Methodology

An *ex-post-facto* research design was followed. The study was conducted in Thiruvananthapuram District of Kerala in six field centres of Vegetable and Fruit Promotion Council, Keralam (VFPCK). Two Self Help Groups were selected from each field centre. A total of 120 Vegetable farmers availing micro credit through VFPCK insurance package were drawn using random sampling technique. Besides 30 experts including extension personnel and bank officials were also selected randomly as respondents.

Findings and Discussions

Constraints was operationalised as those main difficulties or problems which hinder effective servicing and timely repayment of micro credit. Based on review of literature and discussion with extension personnel of VFPCK, a list of constraints were prepared. The respondents were asked to record their extent of agreement or disagreement regarding relevance of these constraints as hindering the micro credit servicing and timely repayment effectively. Also statements related to suggestions to improve present situation were prepared. The scoring was done on a five-point continuum as strongly agree to strongly disagree and score of five to one were given. The total score of each of the statements was calculated and ranking done based on relevance.

The constraints faced in micro credit servicing were ranked by respondents. The most important constraint was formal book keeping procedures which had to be updated. It has been revealed in many studies that NABARD and NGOs supporting micro credit had to make solid efforts to educate and train SHG members for keeping and maintaining records properly for achieving better success in future. Subsidy was an important and attractive benefits for many governmental development programmes in farming sector. But in the case of micro credit, subsidy is absent which is a main constraint. As revealed by many studies in Indian economy, repayment of other debts limits productive utilization of micro credit supplied. Constraints with less relevance were short-term nature of micro credit followed by fixing of season of loan allotment and finally, time of meetings fixed. The number of field staff present, support of bank officials and presence of office building along with presence of non-farming members in SHGs were having medium importance as constraints.

The data were also collected indicating the constraints in micro credit repayment. Labour cost is a major constraint increasing the cost of cultivation and reducing profit. Also market glut resulted as a constraint in adequate marketing facilities leading to low cost for product. Presence of wilful defaulters who are reluctant to pay act as a negative force further reducing repayment levels. Other interesting factors are that political influence is not rated as a major constraint. Also as loans are productive and not consumptive, interest rates is not treated as a constraint. Overdues are also not expected to be written off.

Conclusion

The study revealed the need for more grants and aids to help to build adequate infrastructure. The challenges of micro credit programme are real and change agents will have to struggle for keeping the SHG movement away from subsidy oriented programmes. This can result in a mutually beneficial banking policy with poor becoming more bankable.

REFERENCE

Nagayya, D. (2000). Micro Finance for Self Help Groups, *Kurukshetra* (48): 10-12.

Chapter - 11

Multimedia—An Innovative Tool for Technology Dissemination

Rajula Chandran and *R. Thiagarajan*

Introduction

Modern Information Technology in the place of traditional extension approach is the need of the day in the growing context of globalization of agriculture and allied sectors. The Information Management spectrum extends from the most traditional forms of interpersonal communication and performing arts to sophisticated technologies. Information technology is unfolding and has very high visibility in this sputnik era of communication. The generation and dissemination of information has to be all encompassing, demand driven and responsive to the needs of the intended clients. Computer based multimedia technology is a tool for communicators of all trades and an effective catalyst for change.

The biggest advantage of information technology is that it is far more interactive and personalized that can render services particularly the information as per the needs and requirements of the end users (such a facility makes a favourable impact on adoption and utilization of the improved and innovative techniques in hi-technology farming). The information technology have a convergence of three strands of technologies, viz., computers, micro-electronics and communication. The information technology encompasses such different things as books, print, reprography, telephone network, broadcasting and computers. Information technologies that can be profitably used in transfer of new farm technology are Satellite communication, Geographic Information System, Computer Network, Video, Radio and Reprography.

A relatively new dimension of microcomputer technology entitled "multimedia" possesses the potential to influence learning and knowledge acquisition.

Multimedia technology means different things to different people. It might be a communication tool for some, an artistic medium for others, a teaching and learning tool, or a way to complete a business transaction. Multimedia technology is a tool—not an end in itself. Because multimedia technology is a tool, applications are practically limitless and people from a wide variety of fields are encouraged to learn how to plan and develop multimedia applications in their fields of interest.

What is Multimedia?

The term multimedia describes a number of diverse technologies that allows visual and audio media to be taken and combined in new ways for the purpose of communication. Applications include entertainment, education and advertising. Multimedia often refers to computer technologies. Nearly every PC built today is capable of multimedia because they include a CD-ROM drive, a good sound card and video card. But the term multimedia also describes a number of dedicated media appliances, from digital VCRs and interactive television, to advanced wireless devices, to public television screens. The meaning of multimedia will continue to expand as technology evolves and new applications are invented.

It is a woven combination of text, graphic art, sound, animation, and video elements. When you allow an end user—the viewer of a multimedia project—to control what elements are delivered and when, it is interactive multimedia. When you provide a structure of linked elements through which the user can navigate, interactive multimedia becomes hypermedia. Although the definition of multimedia is a simple one, making it work can be complicated. Not only do you need to understand how to make each multimedia element stand up and dance, but you also need to know how to use multimedia computer tools and technologies to weave them together. The people who weave multimedia are known as multimedia developers.

The software vehicle, the messages, and the content presented on a computer or television screen together constitute a multimedia project. A multimedia project need not be interactive to be called multimedia, users can sit back and watch it just as they do in a movie or the television. In such cases a project is linear, starting at a beginning and running through to an end.

When users are given navigational controls and can wander through the content at will, multimedia becomes non-linear and interactive, and is a very powerful personal gateway to information.

The integration of multimedia technology into the communication environment has the potential to transform an audience from passive recipients of information to active participants in a media rich learning process. The introduction of multimedia or any other computer based information technology is not intended to substitute for a presenter. This new technology is intended to provide to the presenter with a powerful tool that can greatly enhance communication by delivering a multisensory experience. However, a multimedia information kiosk can be designed to provide information to users with or without human intervention.

Where to use Multimedia

Multimedia is appropriate whenever a human interface connects a person to electronic information of any kind. Multimedia enhances traditional text only, computer interfaces and yields measurable benefit by gaining and holding attention and interest; multimedia improves information retention. When properly woven, multimedia can also be profoundly entertaining.

Studies indicate that if you are stimulated with audio, you will have about 20 per cent retention rate, that of audio visual is upto 30 per cent, and in an interactive multimedia presentations where you are really involved, the retention rate is as high as 60 per cent.

The Stages of a Project

Most multimedia projects must be undertaken at stages. Some stages should be completed before other stages begin, and some stages may be skipped or combined. The four basic stages in a multimedia project are :

1. *Planning:* A multimedia project always begins with an idea or a need that you refine by outlining its messages and objectives. Identify how you will make each message and objective work within your authoring system. Before you begin developing plan what writing skills, graphic art, music, video and other multimedia expertise will be required. Develop a creative graphic look and feel, as well as a structure and navigation system that will let the viewer visit the messages and contents.
2. *Costing:* Estimate the time needed to do all elements, and prepare a budget. Work up a short prototype or proof of concept.
3. *Designing and Producing:* Perform each of the planned tasks to create a finished product.
4. *Testing:* Always let your multimedia programs to make sure that they meet the objective of your project, work properly on the intended platforms, and meet the needs of your client or end user.

Prerequisites for a Multimedia Module

Multimedia is often a team effort; artwork is performed by graphic artists, video shoots by video producers, sound editing by audio producers, and programming by programmers. To create multimedia you need to have a real yearning to communicate, because multimedia is creating, essentially, an entirely new syntax for communications. One must have an interest in human psychology because you need to anticipate the brain waves of users. One should adopt a strategy that allows one to prototype and test the interactive design assumptions.

The three basic items for preparing a multimedia module are hardware, software and creativity (good ideas to make multimedia) :

Hardware: There are two most significant platforms for

producing and delivering multimedia projects; the Macintosh OS from Apple and any Intel based IBM-PC clone running Microsoft windows.

Software: Multimedia software tells the hardware what to do. Software is the key, it creates all magic. Multimedia elements are typically sewn together into a project using authoring tools. These software tools are designed to manage individual multimedia elements and provide user interaction. In addition to providing a method for users to interact with the project, most authoring tools also offer facilities for creating and editing text and images and they have extensions to drive videodisk players, videotape players, and other relevant hardware peripherals; sounds and movies are usually created with editing tools dedicated to these media, and then the elements are imported into the authoring system for playback. The hardware and software that govern the limits of what can happen are the multimedia platform or environment.

Creativity: Before beginning a multimedia project you must first develop a sense of its scope and content. Let the project take shape in your head, as you think through the various methods available to get your message across to your target viewers. It is very difficult to learn creativity. Some might say its impossible that one is born with it. But like classical artists who work in paint, marble or bronze, the better you know your medium, the better able you are to express your creativity.

For multimedia, this means you need to know the hardware and software first. Once you are proficient with the hardware and software tools, what then can you build that will look great, sound great and knock the socks of the viewer.

The Multimedia Production Team

High quality interactive multimedia applications are the product of the efforts of a production team. Several software applications are available that enable the average computer user to develop simple and effective presentations. But the production of fine quality high end multimedia applications is usually the work of a team of specialists. Typically this team consists of :

- The Production Manager
- Content Specialist(s)
- Instructional Designer
- Script Writer
- Text Editor
- Multimedia Architect or (program Authoring Specialists)
- Computer Graphic Artist
- Audio and Video Specialists
- Computer Programmer

Quality Multimedia production is a result of effective and synergistic teamwork.

Multimedia Authoring

Multimedia authoring consists of bringing together different elements of multimedia or the building blocks of multimedia such as text, graphics, sound, animation, video etc. into a comprehensive presentation, and if and when needed, incorporation of interactive elements, such as navigation tools. Multimedia authoring requires development of the building blocks, import them and integrate them into a comprehensive and possibly an interactive application.

The application software used for such integration of the elements into a presentation and incorporation of interactivity through navigational controls is called multimedia authoring software or simply authoring tool.

Basic Functions of Multimedia Authoring Tool

1. *Importing:* The tool should have the capability to import the building blocks such as text, graphics, animation, audio files and digitized video files, developed using other software.

2. *Creation:* The authoring tool should have in-built capability to generate or create the building blocks by designing, editing, painting, charting, audio capturing and video processing.
3. *Integration:* The authoring tool should enable integration of the building blocks by sequencing, timing, linking and scripting. It should also be capable of keeping databases for user responses, provide feedback and call on other resources and applications to be run from the multimedia application produced.
4. *Delivery:* The authoring should be capable of developing self-running applications and encryption of the applications.

Selection of Multimedia Authoring Tool

Varied multimedia authoring tools are available in the market. The authoring tools may be classified based on the way by which the software organizes the elements, sequences the events and delivers the application. Which one to choose from them is an important decision. The key consideration is the purpose for which the application is to be developed. So analyzing the goals of proposed multimedia application before the selection of the multimedia authoring tool becomes crucial. The basic considerations are purpose of application, level of expertise available/amenable, time frame for the production, budget and level of cosmetics to be applied. The authoring tools generally used for education purpose includes Macromedia Director, Authorware, Asymetrix Tool Book, Hyper Card, Hyper Studio and Super Card.

Conclusion

Today multimedia in its different forms penetrates a lot of different communication environments. From a technical point of view, multimedia is based on networked communication (e.g.www) as well as digital stored media (e.g. CD-ROM). On the network, multimedia is used for business as well as public information, and the information represent presentations as well as databases.

Furthermore, the information often will be supplemented with possibilities for two way communication, normally through mailing systems, but also as information exchange, e.g., banking, sales etc.

As digital stored multimedia, we find public information systems such as encyclopaedia, as part of exhibitions and as information kiosks. As part of training systems, they are used for simulation as well as learning systems, and their dialectic purpose can be presentation, learning, scripting or reporting. Related to business, multimedia are used for administration (e.g., stock lists), marketing (catalogues) and maintenance (hand books). Finally, multimedia today is often related to leisure through mass cultural products as games as well as more superficial art productions. In accordance to the development of the information technologies, the networked and the digital stored multimedia are converging. But still, the possibilities for more sophisticated solutions are more expanded on the CD-ROM than on the www.

During 1990s, multimedia has been a central medium for communication. But at the same time multimedia is one of the most misused words in the world. Everybody tries to sell everything under this label. Multimedia have been presented as 3D graphics, as simple combinations of text and graphics and as more expanded systems with use of images or video at one side, at the other side as programs without interaction as well as programs weighting interaction from browsing to search techniques, from icon based interaction to symbolic interaction and from use of keyboard to virtual reality. Furthermore, multimedia is connected with stand-alone systems as well as high capacity networks. From a systematic point of view, multimedia today is a dummy-labeling all matters connected with high tech based communication. However, integration of multimedia in the information dissemination allows better comprehension and retention of the subject matter due to the diversity of different media delivery systems.

REFERENCES

Chuck, H. (1997). *Mastering Macromedia Director - 6*, BPB Publications, New Delhi.

John,V.C. and Molina,L. (1998). *Multimedia production, planning and delivery.* Prentice Hall, Inc., New Delhi.

Jose,L. (1998). *Multimedia sound and video*, Prentice Hall, Inc., New Delhi.

Vaughan,T. (2001). *Multimedia: making it work.* Osborne McGraw Hill.

Vijayaraghavan, R. (2003).Key issues in Agricultural Information Technology Management. *IEF Newsletter* 9 (3&4): 5-6.

Chapter - 12

The Internet — A Potential Supplementary Tool for Agricultural Extension

M.V. Sajeev

Introduction

The gap between the information rich and information poor is getting wider and wider. The widening gap has emerged as a big challenge for agricultural extension agents in rural areas. Access to information and improved communication is a crucial requirement for sustainable agricultural development. Modern communication technologies when applied to conditions in rural areas can help improve communication, increase participation, and disseminate information and share knowledge and skills. It is being said that "Cyber Extension" would be the major form of technology dissemination in the near future. However, it is observed that the rural population still has difficulty in accessing crucial information in order to make timely decisions. It is essential that information availability is demand driven rather than supply driven. The challenge is not only to improve the accessibility of communication technology to the rural population but also to improve its relevance to local development. The ongoing IT revolution has opened up huge opportunities for providing access to information as well as to interactive distance learning in rural India. Keeping the above points in view this review paper explains the following aspects :

1. Why Internet?
2. Concept of cyber extension
3. Limitations of traditional extension
4. Advantages of cyber extension

5. The tools of cyber extension
6. The methods of cyber extension
7. Indian experiences and success stories

Why Internet?

The Internet is emerging as a tool with potential to contribute to rural development. It is an electronic infrastructure, a network of networks which opens away to have intense communication and interactions. Internet enables rural communities to receive information and assistance from other development organizations; offers opportunities for two-way communication. It can also support bottom-up articulation of development needs and perceptions, and thus help in reducing the isolation of rural communities. It can facilitate dialogue among communities and with government planners, development agencies, researchers, and technical experts; encourage community participation in decision-making; and help agricultural researchers, technicians, farmers and others in sharing information.

The Concept of Cyber Extension

To understand the concept of cyber extension we have to first understand what cyber space is.

Cyber Space is an imaginary or virtual space of computers connected with each other on networks, across the globe. These computers can access information in the form of text, graphics audio and video and animation files.

Cyber Space thus can be defined as the imaginary space behind the interconnected telecommunications and computer networks, the virtual world.

Cyber Extension thus can be defined as the extension over cyber space.

Limitations of Traditional Extension

1. Traditional Extension is Expensive

Traditional extension requires lot of funds to produce and print messages. The same is the case with training the whole lot of extension personnel.

2. Very Much Time Consuming

Since the number of actors involved in the whole process of transfer of technology is very high, the process takes lot of time and effort on the part of extension machinery.

3. Eroding Quality of Messages while Passing Through Different Layers

Due to the various levels an extension message has to pass through, before it reaches the farmer, communication distortion occurs inevitably and ultimately the quality of message is eroded.

4. Poor Communication Capacity of Existing Extension Systems

This can be attributed to three reasons:

1. Top-down flow of information without interaction.
2. Little use of up-to-date communication technologies.
3. Poor feedback collection and utilization.

5. No Provision for 24×7 Service

Traditional extension can't be availed 24 hours and 7 days a week.

6. Geographical Barriers

Many a times geographical barriers prevent close and frequent interaction between farmer and extension worker.

Advantages of Cyber Extension

1. Provides 24 × 7 service to farmers
2. Can be accessed from any place on the globe
3. Enhanced communication capacities
4. Quality of message is kept intact
5. Very fast service
6. Less costly for farmer compared to traditional extension
7. Provision of value information
8. Direct access to expert advice.

The Tools of Cyber Extension

There are a wide variety of tools used by cyber extension. But the most important ones are two of them :

1. *e-mail*: Electronic mail is the most frequently used tool in cyber extension. It allows us to send and receive text and picture messages from others.
2. *World Wide Web*: The web is the most popular place on the internet. www is an organization of hypertext documents containing text, images, animation, sound, video and increasingly interactive programmes. Websites all around the world can be accessed in www.

The Methods of Cyber Extension

1. Providing interaction among research scientists, extension workers, farmers and other rural people through e-mail.
2. Providing up-to-date news and information services.
3. Question and answer service involving farmers and experts.
4. Creation and maintenance of statistical data base.
5. Providing status of government programs and poverty alleviation schemes.
6. Hosting websites of major institutions in agricultural extension with package of practices.
7. Provide internet access to village people through agriculture and rural development offices.
8. Opening cyber cafes at village level to help extension workers and educated village people to access information from www.
9. Providing video clips of method demonstrations and audio files of local radio station programmes.
10. Providing online classes for farmers on a group basis.

Initiatives in India/Success Stories

In the *Warana Wired Village* Project covering 70 villages in Maharashtra the existing cooperative structure has been used with state of the art infrastructure to allow Internet access to existing cooperative societies. The aim is to provide information to villagers by establishing networked booths in the villages.

The *Information Villages Project* of the M.S. Swaminathan Research Foundation is aimed at bringing the benefits of modern information and communication technologies to rural families in Pondicherry. A Value Addition Centre which is the hub of the information network has been established in Villianur village and four information shops have been established in different villages.

National Dairy Development Board. IT-based machines are being used at milk collection centres, and in cooperatives to measure butterfat content of milk, test the quality of milk, and promptly make payment to the farmers. It has resulted in the removal of incentives to cut the milk by adding water, reduced time for payments from 10 days to less than five minutes, and instilled confidence in farmers in the cooperative set up. All of these factors have helped the milk market to expand.

A CMC pilot project has installed a *Computerized Universal Postal System and a Centralised Accounting and Reporting System* in three post offices in Andhra Pradesh. The technology is designed for rural environments. The systems handle multifunctions within a postal office, reduce errors and waiting time, and provide transparent transactions.

ITC can help empower the knowledge rich but economically poor people. Under the "*Honey-Bee" knowledge network* (of the IIM, Ahmedabad) used to augment grassroots inventors and overcome language, literacy and localism a large number of grass-roots inventions have been identified and documented as short multimedia presentations. Future plans include creating a database of such innovations and making them accessible *via* a wide area network.

One-way video, two-way audio teleconferencing interactive networks have been used for education and training by *Indian Space Research Organisation.* The major application of the network in rural development was for training extension staff from various departments of the state governments. In addition, a large number of women, Panchayati Raj elected officials, primary school teachers, and child development workers spread over large distances have been trained.

Information and communication technologies are an important part of the *Government of Andhra Pradesh's* efforts to improve the efficiency of its administrative offices. AP is the first state in India to design a *statewide computerization program* that will be used in rural areas, at the mandal-level (the administrative unit above the village-level panchayat). There are 1124 mandals in the state. The first software application, is the issuance of certificates pertaining to land holdings, caste, nativity and income across a common counter, without the current delay of 15 to 20 days.

The AP State-wide Area Network (APSWAN) aims to link the State Government's Secretariat with 23 District Headquarters, serving as the backbone for "multi-services" (voice, video and data) that would be used for improved co-ordination between state headquarters and district offices in managing various regulatory developmental, and hazard mitigation programs of the state government. Mandals will be served by this two-way communication, and electronic commerce applications will be developed. The AP Value Added Network Services project hopes to deliver a variety of public services through a large network of information kiosks.

The *Computer-aided Administration of Registration Department (CARD)*, a project of A.P. aims to introduce a transparent system of property valuation, which is easily accessible to citizens.

The *e-choupals* started by *ITC,* which serves 1.8 million farmers in 5 states and is expected to be expanded to other states is providing a host of information like weather conditions, market prices etc. to farmers.

Conclusion

Cyber extension can be used as a supplement in conjugation with existing agricultural extension mechanism. It invariably is not the anecdote for all the existing problems in the field of extension. Also it cannot replace traditional extension in the near future. But it will add more speed and interactivity and in-depth messaging and feedback. Further it will reduce cost, time involved and dependency on so many actors. It will definitely bring quality to Indian agricultural extension and its whole method in the coming decades.

REFERENCES

Sawant, A.G., The changing roles of agricultural extension in agricultural development. Talk delivered at Hans E. Kardel and B.J. Nanjudappa Memorial Endowment Lecture, 28th June 2003, Institution of Agricultural Technologists, Bangalore.

Singh, A.K., *Agricultural Extension: Impact and Assessment*, Agrobios, pp. 31-232.

Manage Cyberaria, accessed Online at *www.manage.gov.in*

Chapter - 13

Raitha Samparka Kendra—An Alternate Demand Driven Extension Approach

M.S. Nataraju, B. Narayanaswamy, K. Narayanagowda and *M. Manjunatha*

Introduction

India's public extension system is considered to be the largest in the world with vertically downward network. It is required to cater to the farming and allied needs of over 90 million farm families, the farmers majority of whom is small and marginal farmers with an average land holding of 1.63 ha. Additionally, such holdings are fragmented and scattered over different agro-climatic zones (Mishra, 1990). In the last five decades, India has experimented with several structures and strategies in agricultural extension. The analysis of National Extension System (Training and Visit) implemented from mid-seventies has brought to surface the following strong and weak points (Dwarakinath, 2000).

Strong Points

1. *Exclusively extension on work*—Supplies and services and regulatory functions are excluded.
2. *Strong links with research*—Master trainers, NAEP and NARP have ensured this feature.
3. *Subject-matter support*—This is provided at both district and sub-divisional levels.
4. *Continuous training*—Both monthly workshops and fortnightly field training sessions are provided for.
5. *Time-bound operations*—Predetermined, periodic field visits and supervision are built into the activities.

6. *Single line of command*—Technological and administrative control is vested in the Department of Agriculture.
7. *Unified extension work*—All commodity schemes, special programmes and farm-based production activities are suggested to be brought together to be implemented by one agency.
8. *Adherence to management principles*—Specified jurisdiction, manageable span of control, periodicity of visits and adequate training are supposed to be based on the precepts of sound management.

Weak Points

1. The focus is too narrow limiting to crop production technologies.
2. Stale and inadequate technology development.
3. Excessive emphasis 'talking' type of extension instead of 'doing' type of extension work.
4. Top-down communication of information.
5. The whole operation is controlled by managers and scientists and at no stage it is a partnership with the farmers.
6. Low status of village level workers.
7. Much efforts to improve extension services has concentrated upon the efficient delivery of messages rather than upon the relevance and effectiveness of the message themselves.
8. Farmers are supplied with standardized advice, regardless of local circumstances or individual needs.
9. Lack of sufficient number of efficient extension workers.
10. Lack of co-ordination between research-extension and user system and development departments.

It is expected that in the coming years farming community will gradually differentiate themselves into two distinct categories. One will be venturesome in aiming at domestic and export markets and another is large segment of small and marginal farmers who depend on general extension services. To meet these new challenges and requirements in the era of globalization a new strategy called Raitha Mitra Yojane is being implemented in Karnataka.

It is a new demand driven extension strategy conceptualized during 2000 to meet new challenges like :

- Yield improvement with better use of resources
- To bring in the concept of whole farm development
- More attention to rainfed farming and better management of irrigation resources
- Tie up with both private and public input agencies
- Providing information on production options and better technology
- Providing facility of primary testing of seeds, soil and other inputs.

Raitha Samparka Kendra

The agricultural centre opened under this new programme at Hobli are called 'Raitha Samparka Kendra'. RSK will be located in farms adjacent to a district main road and are located in such a way that they are not more than three km. from Hobli Headquarters. Near the vicinity of RSK there should be agricultural lands. RSK would have at least 4 acre of land so that it can promote on site demonstration of seeds and other inputs. It must have minimum 600 sq. ft of building space initially to accommodate information room, primary testing laboratory, input provision room, office space for AA's and AO's apart from facilities for crop demonstration.

Objectives of RSKs

1. Providing Updated Information on Crop Production Practices and Option and Market Trends

The extension wing of department of agriculture has undertaken the work of imparting information on production practices and popularizing various inputs. The AA's and AO's will attend the request made by farmers apart from routine visits. These RSKs are provided with Internet facility, so that they can access information on production options and practices and market trends anywhere from the world of agriculture. Technical persons are also going to visit the fields by farmers request.

2. To Facilitate on Site Provision of Agricultural Inputs like Fertilizers, Plant Protection, Chemicals and Seeds etc.

The farmers are forced to travel to the HQ to purchase Plant Protection chemicals and seeds as the inputs are not available at the village point. If the private sectors and public agencies like KSSC and KAIC are given some storage space on rental basis then probably they would stock and crucial inputs at the Kendra level itself. By this, farmers would be in a position to get access to input at the level of RSKs and Kendra would be earning some revenue for further development.

3. To Facilitate on Site Provision of Primary Testing Facilities like Seed Germination, Quality Testing and Soil Tests etc.

Seeds and fertilizers are two crucial inputs in agriculture. As the certified seeds are available to an extent of 3 per cent of the requirement so primary testing of seeds has promoted in Kendra. The fertilizers are becoming costly because of withdrawal of subsidies and rationalized application is required to reduce the costs. Need based application of fertilizers require soil testing. This facility of soil testing is promoted in Kendra. Soil sample submitted will be analyzed with the help of mobile soil testing lab and they will be charged as to cover the cost of testing.

4. To Provide for Demonstration of both Public and Private Seed Materials and Inputs at Kendra Level

Both universities of agricultural sciences and many private companies are engaging in promoting new high yielding varieties, field demonstration is a very important way of demonstrating new varieties. The RSKs will provide demonstration space within its premises on payment of ground rent for four months to public agencies like KSSC and private seed companies for their traits in

Hobli. This will enable farmers to compare the performance of various varieties under common conditions in one place. Secondly, planting of the popular cultivars during every season will act as a reference control plot for the whole Hobli during crop season even demonstration on new inputs like granulated fertilizers or bio-fertilizers, neem based plant protection chemicals etc. can also be done. The comparative demonstrations will give an opportunity to farmers, researchers, seed companies and government to get an opportunity to evaluate the available technologies.

5. ***To Provide a Forum for Interphase with Public and Private Technologies and Inputs***

There are number of new technologies including new varieties are coming into market. RSKs are having one information centre where government and private sectors can put up information on their new technologies. But private agencies should pay for wall space and exhibition space for their poster or samples. This will ensure more competitive technology is exhibited and material will be replaced as they become old.

Financial Stability of RSKs

The salary of the AAs and AOs, the cost of various crop demonstration activities are already provided in the budget. User charges will be collected at certain rates approved by government from various private companies involved in agricultural sectors and also from the farmer's revenue generated in this manner will be utilized for the further development of extension system. So there may not be any additional recurring expenditure on account of opening of RSKs.

Staff Composition of RSKs

Each RSK will have two AOs and five AAs. While one AA and AAO remains at the RSK. The other four AAs and AOs will go on field visit in their respective jurisdiction as and when demanded by the farmers. They would attend both extension work as well as other assigned work like field experiments.

Extension Functions and Other Services of RSKs

1. Integrating with public and private supply agencies is to be organized at RSKs for farmers.
2. Giving information about improved agricultural implements.
3. RSKs organizes programmes for popularizing agricultural credit and crop insurance.
4. Selection of beneficiaries for different schemes of government and helping them to take benefits to increase production.
5. Demonstration of public and private sector's improved technology is undertaken by RSKs on payment basis.
6. Private sector service providers can use RSK for meeting and other activities with farmers.
7. They help to give publicity to the private and public agriculture agencies programme about farming.
8. They test the seeds given by the farmers.
9. They test the soil on request by farmers and also based on the soil test results they recommend the fertilizer to be applied to their field.
10. Based on farming problems of the farmers visiting RSKs they recommend scientific technical suggestions to them.
11. Giving training to the students studying the discipline of agriculture.
12. RSKs should act as a discussion centres for soiving the problems faced by the farmers, farm youth and farm women.

Conclusion

Raitha Samparka Kendras, which is a demand-driven alternate extension strategy, takes care of most of the requirements of the farmers. It also helps in maintaining better rapport between farmers and officials of Department of Agriculture and one of the self-sustained system compared to the previously existing systems. By observing the number of farmers visited at RSKs during the previous years, it may be inferred that the response towards RSKs is encouraging. It may be noticed that RSKs facilitate on site provision of agricultural and other facilities. On the positive side of it they are attaining to generate funds of their own. Hence, the concept of RSKs looks unique in present context of dissemination of agricultural technologies that is going to be timely and needful extension strategy for the farming community.

REFERENCES

Anonymous, (2000), Agricultural Extension. In : *www.raitha mitra.kar mic in*

Dwarakinath, R., (2000), *Extension Implications of a Changing Agriculture in India*, IAT, Bangalore.

Mishra, D.C., (1990), *New Directions in Extension Training. A Conceptual Framework.* Directorate of Extension, Ministry of Agriculture, New Delhi.

Chapter - 14

Integrated Farming System Demonstrations

An Innovative Extension Approach

V. Veerabhadraiah and *G.N. Nagaraja*

Introduction

The most spectacular achievement in India is the self-reliance in food grain production. But our food security is to be threatened in the near future due to high population pressure. According to a recent estimate, the foodgrain requirement of India would be 380 million tonnes (mt) and 450 mt in the years 2025 and 2050, respectively, with the present rate of population growth. Use of high yielding varieties, extension of irrigation facilities and high application of fertilizers and pesticides contributed to radical enhancement of foodgrain production in the mid-60s and 70s of the last century which is known as green revolution. The green revolution concentrated only on single enterprise, crop (more specifically wheat, rice and maize). Although, later on, development in other enterprises like dairy, poultry, piggery, fishery etc. took place, in an isolated manner. The only wayout is to diversify the economic activities in an interdependent and integrated manner at micro level within the available resources. In other words, it is the integration of two or more enterprises like crop, dairy, piggery, fishery, poultry, bee-keeping etc. for each farm according to the availability of resources to sustain and satisfy the necessities of the farmer. The system thus developed will be termed as integrated farming system approach. It is possible even for a resource poor farmer and for him integrated system is more important to maximize the profit through proper utilization of scarce resources in a farming system approach.

Integrated Farming System Demonstration (IFSD) is an innovative and unique approach to promote integrated land use and animal management technologies as well as resource management capabilities among the farmers, more particularly small and marginal farmers of rural populace. It is a micro approach and the entire farm of an individual farmer is considered as a unit. This approach calls for concerted educational efforts with farmers with primary focus of maximizing the net income of farmers over a period of time. This is closely related to realistic planning of farms of selected farmers in order to help to generate maximum family employment and to get sustained stable net income. Such planning and implementation requires identification of potential for development, needs, interests and capabilities of farmers, availability of resources and also training them in farm management techniques. The different enterprises not only compete for scarce resources such as land, labour and capital on the farm but also exhibit inter-dependence due to supplementary or complementary relationship. This also helps in recycling of farm wastes and residues without polluting the environment and degrading the resource base.

Further, Integrated Farming System (IFS) gained priority with a view to augmenting the income of farmers. It assumes great importance to evolve sound management of farm resources to enhance farm productivity, reduce the degradation of environmental quality, improve the quality of life of small and marginal farmers and above all to maintain sustainability in farm production and productivity.

In short, the overall objective of the approach is to evolve technically feasible, economically viable, environmentally sound and socially acceptable farming system models by integrating crops with appropriate supplementary and complementary enterprises for rainfed and irrigated areas. This helps in generating sustained stable income and higher employment from the entire farm.

Characteristics of the Integrated Farming System Demonstration

1. *It is farmer-oriented* - IFS approach targets small and marginal farmers as the clients of agricultural research and technology

development. Consequently, its basic objective is to make technology generation more relevant to their goals, needs and priorities.

Several mechanisms are commonly employed in the approach to attain this objective. Farmers are made an integrated participant in the research process. The existing farming system is studied before proposing improved technologies. The Farm Management Extension Specialist (Social Scientist) collaborates with other agricultural scientists on the analysis of the existing system, constraint analysis, and development of alternative technologies.

2. *It is a system-oriented* - The IFS demonstration views the whole farm and farm household and the interrelation and interaction between farm enterprises. This is necessary for understanding the complexity and functioning of small and marginal farm agriculture and for the diagnosis of farm crop production problems.
3. *It employs a problem-solving strategy* - It is essentially operational research, which first identifies technical, biological and socio-economic constraints at the farm-level for major types of farming systems. It endeavours to develop technologies that are meant to overcome those constraints.
4. *It is interdisciplinary and multidisciplinary* - Collaboration among agricultural scientists of various disciplines and farm management specialist is needed to understand the conditions under which small and marginal farmers operate. This is aimed at accurately diagnosing constraints and transferring improved technologies suitable to those conditions.
5. *It tests technology in on-farm trials* - On farm experimentation establishes the context for collaboration between farmers and researchers, and encourages a deeper understanding of the farming systems among researchers. It also provides for the evaluation of new technologies under the environmental and management conditions in which it will be used.
6. *It provides feedback from farmers* - IFS provides feedback on farmers' goals, needs, priorities and criteria for evaluating new

technologies to zonal/agricultural research stations and regional policy-makers.

Objectives

The specific objectives of this approach are:

1. To encourage farmers to take up improvement in all the crops grown by them in different seasons by demonstrating new agricultural technologies,
2. To assist farmers in introducing other subsidiary enterprises like dairy, poultry, fisheries, sericulture, piggery etc., to utilize all the available resources more efficiently,
3. To educate farmers to make them account conscious,
4. To help farmers improve their standard of living by working with them over a period of time, and
5. To develop Integrated Farming System Units as centres of agricultural development in the local areas so that other farmers will be able to use them as sources of new agricultural technology leading to rapid spread of farm innovations.

Important Steps

1. Collection of benchmark information on the existing cropping pattern and other enterprises, farm resources and potentialities and also the socio-economic condition of the family.
2. Preparation of appropriate and feasible complete farm plan based on the type of land, irrigation facilities and other such physical as well as financial resources at their command.
3. Farm plans include introduction of new crop varieties/ hybrids, practices, methods, and technologies along with related enterprises.
4. Farm plans guide the demonstrators to follow dry farming practices for soil and water conservation.

5. Helping farmers to secure required financial assistance from financial institutions for purchase of inputs, animals, farm machinery and equipment etc.
6. Working with farmers for a period of three or four years to achieve stable net income at regular periodic intervals.
7. Involvement of selected farmers in planning, implementation and evaluation of the programme.
8. Using IFSD as centres to educate other farmers through educational activities like field visits, field days and interaction sessions etc.
9. Non-provision of subsidies to farmers for conducting of demonstrations.

Impact of IFSDs on Socio-economic Status of Farmers

IFSDs are conducted in all the four Extension Education Units of UAS, Bangalore. In Extension Education Unit, Bangalore covering four districts and 11 taluks, 22 IFSDs have been established during 2002-03. In all, 630 IFSDs were established since 1970 when the approach was adopted. The impact of these demonstrations on farmers is as follows :

1. Consequent to adoption of Integrated Farming System approach, there was a remarkable increase in the net farm income ranging from 70 per cent to 1320 per cent in the demonstration farms due to adoption of technologies in various enterprises.
2. Through the use of HYV/hybrids and application of both high and low cost technologies, the farmers were able to increase productivity and production of crops. The yield increase in different crops ranged from 7 per cent to 43 per cent.
3. Farmers used organic manures, crop residues, animal waste which helped to improve soil health and conservation of soil moisture thereby reducing environmental pollution and enabled to get sustained agricultural production.
4. IFSD farmers can get the cash receipts round the year

through subsidiary enterprises by way of selling milk, meat, eggs, cocoons, etc., which induces the farmers to adopt new technologies and to meet domestic requirement.

5. Appropriate combination of crop and livestock enterprises to take advantage of complementary and supplementary relationship between them has helped in providing more family employment opportunities. This also helps to utilize all the locally available resources efficiently and economically to get sustained stable income from the entire farm.
6. In this approach, every piece of land area was efficiently utilized by growing of perennial fodder crops all along the border of the farms and the places where crops cannot be grown to meet the fodder demand of the animals.
7. Integrated pest management followed in cabbage, red gram and tomato crops was found to reduce pests' problem very effectively and the technology has been spread to many other farmers in the village and around.
8. Timely application of fertilizers and adoption of appropriate plant protection measures in IFSD helped farmers to get maximum yield. The same technology was rapidly diffused to many farmers around IFS farms.
9. Systematic establishment of nutrition garden helped farm women to understand the importance of seasonal fruits and vegetables in their daily diet and family nutrition in addition to earn some cash income by sale of surplus products.
10. Tackling of infertility problem in animals helped to boost the animal and milk production and appreciation in the value of animals.
11. In IFS, by way of effective recycling of organic waste (cow dung and sericulture waste) available in the system can be utilized to generate biogas. From this anticipated energy problem can be solved.
12. Integrated Farming System demonstrators served as centres for dissemination of new farm information to other farmers in the village and around.

13. In addition to new technologies, seed material of some new HYVs of crops moved to other farmers through IFSD farms.
14. IFSDs motivated and influenced many farmers to adopt new agricultural technologies in their farms.
15. Several extension educational activities like group discussion meetings, interaction sessions, field days and field visits were organized around IFS demonstrations and transferred a good number of useful and applicable technologies to a large number of farmers and extension workers.
16. The other indirect changes observed due to IFSDs include changes in food habits, health care, nutritional status, clothing, material possession (agriculture and non-agriculture), saving pattern and shift in borrowing from non-institutional to institutional loan.

Conclusion

The integrated farming system approach has helped the farmers to re-organize the farm business through reallocation of available resources to get sustained stable returns from the entire farm. There is a great need for this approach to improve managerial ability of the farmers. Further, the small and marginal farmers would need support in training, demonstration, extension services, credit, market support as well as supply of inputs must be as part of the integrated farming system. Hence, it is high time for the scientists, administrators and planners to think on such lines and provide adequate facilities and encouragement to the farmers to go ahead with the integrated farming systems with a sense of total commitment. This will not only enhance their income and living standard of their families but also reduce hunger and poverty from the society. Therefore, research initiatives have to be undertaken for developing site-specific farming systems incorporating two or more feasible and appropriate enterprises for different categories of farmers based on their resource endowment.

Chapter - 15

Adaptability and Proliferation of Techno-Interventions under Institution Village Linkage Programme in a Coastal Village

R. Sathiadhas, Sheela Immanuel, Sindhu Sadanandan and *K.N. Jayan*

Introduction

Agriculture, animal husbandry and fisheries provide maximum employment in the primary sector and form the major source of income and livelihood security of about 70 per cent of population in India. After the introduction of liberalisation policy and giving thrust for the adoption of improved technologies, there has been a spectacular increase in production in all these segments and the country has witnessed a rapid structural change. So far in the process of technology generation, upgradation and refinement farmer's active participation was repeatedly overlooked particularly in case of resource poor farmers, which often led to lopsided development. The Institution Village Linkage Programme (IVLP) model executed in the coastal agro-ecosystem of Elamkunnapuzha village of Kerala state, under Central Marine Fisheries Research Institute (CMFRI) is the latest and the most farmer oriented extension approach initiated by the Indian Council of Agricultural Research (ICAR) under the National Agricultural Technology Project (NATP), funded by World Bank for joint diagnostic studies of production system and to expand adaptive research by means of location specific assessment and refinement of technologies. Fisheries play a predominant role as a major occupation of the vast majority of people in Elamkunnapuzha village. In addition to the fisheries based livelihood, livestock management along with agricultural activities form the other supplementary avocations. This supplementary job assumes enormous importance due to the seasonal nature of fisheries. Hence in order to bring about an integrated whole village development a range of techno-interventions generated by National Agricultural Research System (NARS) in agriculture, livestock and fisheries were incorporated. The present paper deals with the adaptability of IVLP model implemented by CMFRI at Elamkunnapuzha with the following specific objectives :

1. To assess the coastal agro-ecology and production system perspectives of various technologies of the village.
2. To identify the problems faced by fisherfolk in the village.
3. To introduce improvements in the existing production systems through better scientific and management practices to enhance productivity without endangering the stability and sustainability of the environment.
4. To examine the impact of scientific production technologies in the coastal agro-eco-system.

Methodology

The present study was conducted in Elamkunnapuzha village of Vypeen Island in Ernakulam district. The sample constitute 300 selected IVLP farmers participated in the techno-interventions. In order to analyse the spatial pattern of the village, problems faced by the farmers, various Participatory Rural Appraisal (PRA) techniques were used in the study. An interview schedule was prepared for collecting data from the respondents. The schedule was pre-tested and necessary modifications were made. Simple statistical tools were used to analyse the data and interpretation of results.

Participatory Rural Appraisal (PRA) Techniques

In order to analyse the spatial pattern of the village, various PRA techniques were adhered. PRA came as a research approach by late 1980s. The term PRA was first used in 1988. It is a continuation of Rapid Rural Appraisal (RRA). PRA forms one of the major components of extension methodologies. It is rapid, cost-effective and an accurate method of research and learning. PRA is used to describe new approaches and methods in which rural people themselves do much of the investigation, presentation, analysis, planning and evaluation of findings.

It is important to remember that no technology helps anyone unless the farmers understand the problems they are facing, appreciate the value of the new technology, and start to use it. Most of the development programmes attempted in the past focused emphasis for technology transfer without considering the need of the end users/farmers. Even though many efforts have been put forth, this approach has resulted in many failures. Many of the technologies promoted to the farmers have been either rejected or *partially* adopted, not only because farmers failed to understand the technology developed by the scientists and promoted by the extensionists, but more often, because the researchers and extensionists failed to understand the farmers for whom the technology was supposedly intended. This *communication gap* has led to many *disappointing* programme experiences. In the conventional approach data is collected and taken to the research institution to be analyzed and used as a base for formulating programmes. In most cases, the farmers being interviewed are not aware of the purpose of the research. Partly because of these flaws in conventional approaches, there has been a recent rapid expansion in participatory methods and approaches. Methods are being used not just for local people to inform outsiders, but also for people's own analysis of their own conditions (Chambers, 1992, 1992 a ; Pretty & Chambers, 1993; Pretty, 1995).

PRA Techniques Used in the Study

(a) *Transect walks and direct observation* : These are systematic walks with key informants through the area of interest, observing, asking, listening, looking, and seeking problems and solutions. The findings were mapped on a transect diagram. This walk gave the IVLP team members an insight about the surprising topography, natural resources, cropping pattern, soil types, natural vegetation and local practices. It has been instructive for many professionals to realize how much they do not see or do not think to ask about. This also successfully helped the researchers to have a graphic representation of places where unexploited or underexploited potential exists in the community.

(b) *Wealth rankings and Livelihood analysis* : Wealth ranking is used to classify households according to relative wealth or well-being. Here the ranking was conducted directly on a social map. Villagers were asked to indicate on the houses the relative wealth classes. Individual assets such as land ownership, animals, and tools were also marked for each household. This type of wealth rankings produce a baseline against which future intervention impact can be measured; providing a sample frame to cross-check the relative wealth of informants who have been or will be interviewed; and producing local indicators of welfare. Through livelihood analysis it was found that fishing and related activities provide the major source of income accounting about 55 per cent followed by coconut 20 per cent, paddy and vegetables 10 per cent, milk and meat 10 per cent and others 5 per cent.

(c) *Diagramming and visual construction* : In formal surveys, information is taken by interviewers, who transform what people say into their own language. But here diagramming by local people was introduced to give them a share in the creation and analysis of knowledge, providing a focus for dialogue, which can be sequentially modified and extended. Local categories, criteria, and symbols are used during diagramming. This is used because rather than answering questions, which are directed by the values of outside professional, local people can explore creatively their own versions of their worlds. Visualizations, therefore, helped to balance dialogue and increase the depth and intensity of discussion.

(d) *Participatory mapping and modelling* : This involved constructing, on the paper, maps using coloured pens and paper. Maps like *resource maps* of catchments, villages, forests, fields, farms, home gardens, *social maps* of residential areas of a village, wealth rankings and household assets surveys, *health maps*, where the health status of each family member was shown using coloured stickers and other markers, impact monitoring maps where villagers recorded pest incidence, input usage, weed distribution, soil quality, and so forth. Some of the most illuminating maps combined historical views with those of the present.

(f) *Seasonal calendars and activity profiles* : Seasonal constraints and opportunities were diagrammed month by month throughout the year. Farmers used pieces of stick, drawn histograms in the dust to represent relative quantities and patterns of rainfall, soil moisture, crops, labour, food consumption, illnesses, prices, animal fodder, fuel, migration, pests, income, expenditure, and so on.

(g) *Time lines* : Historical analyses found to be a good icebreaker for field exercises and include detailed accounts of the past, of how things have changed, particularly focusing on relationships and trends. These include technology histories and review, crop histories and biographies, livestock breed histories, labour availability. Folklore and songs are valuable resources for exploring history.

(h) *Venn and network diagrams* : Venn diagrams involved the use of circles of paper and card to represent people, groups, and institutions. Farmers were given instruction and were asked to arrange the cards to represent real linkages and distance between individuals and institutions. Overlap indicated flows of information and distance on the diagram represented lack of contact.

(i) *Matrix scoring* : This method is for learning about local people's categories, criteria, choices and priorities. Matrix scoring takes criteria for the rows in a matrix and items for columns, and people complete the boxes row by row. Through matrix scoring it was found that fish is mostly preferred because it is locally feasible, profitable and has got market value. Even though crab is profitable, it commands a comparatively lesser preference, due to lack of seed availability.

Micro Farming Situations

Agro-eco-system analysis and resource inventory of the village indicated five micro-farming situations, such as, (1) Tide-fed brackish water system, (2) Open sea-based coastal agro-eco-system, (3) Homestead animal husbandry and poultry farming system, (4) Rain-fed agri-horticulture system and (5) Low-lying seasonal paddy (*pokkali*) lands. Details of the prevailing micro farming situations of Elamkunnapuzha village are given below :

(1) Tide-fed Brackish Water System

This system includes the rivulets having inlets from the sea or a brackish water locale. In Elamkunnapuzha village, the bar mouth and the major rivulet, *Kuzhali* (connecting *Bandhar-Canal* at Malippuram North) and originating from the Cochin estuary are connected to the minor rivulets. The mangrove aside the channels and the fishing ponds act as the breeding ground for shrimps as well as finfishes throughout the year. These fish farms have the advantage of convenient exchange of water from the sea to the ponds during high tide and back to the sea during low tides. The tidal flow into the fishponds usually takes place "before sunrise" and outflow of water (low tide) happens by nearing sunset time. The exchange of water and the entry of fish resources are controlled through fabricated sluices, locally known as *thoompu*. Such exchange enables the culturing of shrimps and finfishes in the ponds.

(2) Open Sea-Based Coastal Eco-system

The village has a coastal stretch of around 5 kms. Fishing is the main activity in the coastal eco-system throughout the year except for the monsoon season. Usually, trawlers, purse seiners, gill-netters, ring-seiners and other types of motorised and non-motorised fishing units with different craft and gear combinations operate in this coastal eco-system. The village has got two major fish landing centres and a seafood export firm. Secondary to fishing, the major commercial activities included in the system are fish marketing (including auctioning), sub contracting of fish vending to the women labourers, peeling and sun drying of fish. With the advent of the People's Plan Campaign and the consequent upraise of decentralised governance have given a new strength to women through Self-Help-Groups (SHGs). The SHGs have contributed an organisational cohesiveness to the women in the village and around 70 per cent of the SHGs are engaged in fishery related ventures, especially in the processing and marketing of value added fish products.

(3) Homestead Animal Husbandry and Poultry Farming Systems

Homestead denotes those areas around the house, which are used for farming and other income generating activities. Organised livestock farming is absent in the area, but keeping few birds and rearing one or two goats are popular practice. Very few people prefer to keep cows for milk production. Ducks, pigs and rabbits are also reared on a smaller scale. Poultry forms an integral part of homestead farming. Many farmers feel that this type of "backyard" poultry keeping is profitable since no inputs or labour is utilised. But really it is not so. Poor producing ability of the birds, small size of the egg, high mortality due to diseases and parasitic infestation make backyard poultry farming less remunerative and uneconomic. Adopting scientific strategies among the livestock based farming systems can significantly increase the profits. Goat farming is one of the popular avocations of the area. The high price for goat meat and ready local demand has prompted many to take up goat rearing. The milk of goat has only limited market demand and rarely been sold out. It is mostly used for feeding kids. High humidity and waterlogged conditions favour parasitic infestations. Limited availability of fodder grasses is also a constraint for goat production. With all the limitations there are good prospects for goat rearing owing to ease of maintenance within the small-holdings and the market demand it commands. Above all, the animal fits well in the homestead-farming region of the coastal eco-system. Proper parasitic control and supplementary feeding coupled with timely disease management can improve the production.

Dairy farming suffers from socio-economic bottlenecks like deficiency of capital, high cost of cattle feed, lack of know-how about management of high producing animals, inadequacy of land for fodder production, poor disease control and scarcity of natural grass and water. High salinity of soil and water makes the problems worse. Usually, the farmers prefer *Uppootha*, a mangrove leaf, as feed. A great deal of work has to be done to make dairying a viable proposition in the coastal agro-eco-system, starting from marketing of milk to disposal of dung and shed washings.

(4) Rain-fed Agri-horticulture System on Embankments and Homesteads

(a) *Coconut based system under rainfed conditions:* Coconut is the dominant crop of embankments and homesteads. Vegetables, tuber crops and banana are grown as inter-crops. The high soil salinity in the region is found to be ideal for coconut plantations. Fertilisation based on scientific management practices is lacking for the plantations and this cultivation requires Integrated Nutrient Management (INM) for enhancing production.

(b) *Vegetable based system on embankments*: Vegetables like amaranthus, snake gourd, cucumber, bittergourd, ridge gourd and cowpea are grown on commercial scale in the embankments of ponds mainly during the monsoon season. The nutrient-rich pond-bed soil is used for fertilising vegetable farms on the embankments. Farmwomen can very well manage these activities, as it requires less effort. Even, family as a whole can take up this activity.

(c) *Banana-based System under Rain-fed Low Lands*: Banana is raised in lowland under rain-fed conditions. Most of the farmers cultivate banana on commercial basis. The availability of fresh water for irrigating banana is relatively good in the interior places of the village.

(d) *Low-lying Seasonal Paddy (Pokkali) Lands:* The farming of paddy as well as shrimp in the paddy fields is done mainly on group farming basis. Paddy is cultivated during June-September exclusively depending on rain and usually mono-cropping pattern is followed. Harvesting of paddy is done within 110-130 days from the time of sowing. Shrimp farming is the alternative crop in the *pokkali* lands after paddy cultivation. Leasing of paddy lands for shrimp farming during November/December-March/April is common in the region.

Problems Identified and Remedial Measures

Need assessment of the target group is one of the most important steps in planning and developing extension programmes. Need assessment is a systematic process for establishing priorities and making decisions regarding programme planning, development and operations. In this study farmer's needs were assessed using informal methods like group meetings and social gathering. This conversation provided insights into the problem faced by the individuals and an idea about what could be done to address the problems. Following are the major problems identified in the village.

(a) ***Fisheries based Enterprises***

1. Lack of knowledge about improved culture systems
2. Unscientific stocking density
3. Inadequate financial support
4. Social problems like poaching, poisoning etc.
5. Poor water exchange
6. Poor water quality
7. Lack of availability of feed
8. Lack of availability of quality seed

(b) ***Livestock based Enterprises***

1. Lack of capital
2. High cost of feed
3. Lack of nutritious feed
4. Lack of knowledge about fodder production
5. Inadequate land for fodder cultivation
6. Lack of knowledge about management of high yielding animals
7. Disease occurrence
8. Water logged conditions

(c) ***Agriculture based avocations***

1. Soil salinity
2. Lack of finance
3. Lack of institutional support
4. Lack of knowledge about improved package of practices
5. Lack of awareness of new varieties
6. High cost of labour
7. Unfavourable attitude towards scientific cultivation

Based on the problems identified, need based and location specific techno-interventions were implemented with the active participation of farmers. Some of the successful interventions are shown in Table 15.1 to Table 15.4 :

Conclusion

The new extension approach of IVLP introduced by ICAR is highly efficient in transfer of technology in terms of its adoption level, proliferation and spread effect having far reaching implications on productivity, profitability and livelihood security of rural household inhabited under different agro-eco-system.

The interventions, which addressed the aforesaid problems in the coastal agro-system convinced the farmers about the need for proper water exchange, uniform stocking size and eradication of predators prior to stocking and enabled the farmers to use the

Table 15.1. Techno interventions evolved as potential solutions

Fisheries based interventions	*Livestock based interventions*	*Agriculture based interventions*
1. Monoculture of uniform sized juvenile crabs	1. De-worming, micronutrient supplementation and prophylactic immunization on the productive performance of dairy cows	1. Improved variety of Amaranthus, Bittergourd, Snakegourd
2. Polyculture of finfish	2. De-worming, micronutrient supplementation and disease control on the performance of goat	2. Tissue culture of Dwarf-cavendish
3. Integrated farming of fish and poultry	3. Gramalakshmi breed in comparison to country birds	3. Nutrient management practices in coconut plantations
4. Rack drying method of fish drying	4. Duck rearing practices	4. Improved variety of ridgegourd, salad-cucumber and vegetable cowpea
5. Monoculture of *M.cephalus*	5. Broiler rabbit farming in homesteads	
6. Crab fattening	6. Fodder cultivation on unutilised marshy lands	

Results

Table 15.2. Impact of fisheries based interventions

S.No.	Name of the intervention and Farmers practice (FP)	Treatment (TM)	Impact: Net returns (Rs/ha)/unit		Impact: B.C. Ratio	
			Before (FP)	After (TM)	Before (FP)	After (TM)
1.	*Mono-culture of uniform sized juvenile crabs* Uneven stocking of crab seeds along with other fish varieties	Mono-culture of uniform sized juvenile crabs [stocking rate: 4800nos/ha, size: 150-200g]	4,03,800	7,43,568	1.36:1	2.10:1
2.	*Polyculture of finfish* Improper stocking of species and eradication of predators	Eradication of predators using mahua oil cake, uniform stocking of *C.chanos* and *M.cephalus*	2,53,763	4,43,352	0.96:1	1.64:1
3.	*Integrated farming of fish and poultry* Uneconomic farming practices	Fertilization of ponds using poultry droppings, and stocking of *M.cephalus* and *C.chanos*	2,65,313	5,51,740	0.93:1	1.85:1
4.	*Rack drying method of fish drying* (One unit-I5 members/unit) Traditional sundrying of fish	Rack drying over traditional method of sun drying.	3,941	5,901	1.37:1	1.98:1
5.	*Mono-culture of M.cephalus* Improper stocking of species and no proper eradication	Eradication of predators using mahua oil cake and stocking of seeds at the rate of 20,000nos/ha	1,17,875	2,75,750	0.64:1	1.45:1
6.	*Crab fattening* Farming of water crabs	Stocking of water crabs and hardening them using better management practices	3,53,000	6,00,000	1.24:1	2.50:1

Table 15.3. Impact of livestock based interventions

S.No.	*Name of the intervention and Farmers practice (FP)*	*Treatment (TM)*	*Impact*			
			Net returns (Rs/ha)/unit		*B.C. Ratio*	
			Before (FP)	*After (TM)*	*Before (FP)*	*After (TM)*
1.	*De-worming, micronutrient supplementation and prophylactic immunization on the productive performance of dairy cows* Inadequate and improper medication, deworming, vaccinations against infectious diseases	Mineral/Vitamin supplementation, Vaccination against Foot and Mouth Disease	41.75	55.50	1.12:1	1.42:1
2.	*De-worming, micronutrient supplementation and disease control on the performance of goat* Traditional practice, indigenous medicines	De-worming, Mineral/Vitamin supplementa tion, Vaccination against Foot and Mouth Disease	1.12	1.85	1.00:1	1.39:1
3.	*Gramalakshmi breed in comparison to country birds* Grow country birds of low productivity and long broodiness. No practice of controlling parasites or prophylactic vaccination	New breed (*Gramalakshmi*), Mineral/ Vitamin supplementation, Vaccination (30 Birds)	2.42	5.96	0.61:1	1.12:1

Table 15.3 *(Contd..)*

4	*Duck rearing practices*					
	Rearing of low productive country birds	Ducklings of high genetic potential (*Kuttandan*), better feeding	7.70	12.00	1.05:1	1.33:1
5.	*Broiler rabbit farming in homesteads*					
	Grow local breeds of poor productivity	Introduction of hybrid breed of rabbit (*Grey giant*)	2.95	5.75	1.1:1	1.46:1
6.	*Fodder cultivation on unutilised marshy lands*					
	Paddy straw and Uppootha	Introduction of paragrass	—	97.2	—	1.16:1

Table 15.4 : Impact of agriculture based interventions

S.No.	Name of the intervention and Farmers practice (FP)	Treatment (TM)	Impact			
			Net returns (Rs/ha)/unit		B.C. Ratio	
			Before (FP)	After (TM)	Before (FP)	After (TM)
1.	*Improved variety of Amaranthus, Bittergourd, Snakegourd*					
	Traditional farming of local variety and uneven spacing pattern and seed rate	Cultivation of improved variety and recommended spacing and seed rate (a)Amaranthus:Kannara local, 30cm×20cm,8 g/cent (b) Bittergourd: Preethi, 2m×2m, 20g/cent (c) Snakegourd: Kaumudi, 2m×2m, 16g/cent	(a) 40,500 (b) 60,625 (c) 38,625	(a) 54,275 (b) 1,28,325 (c) 55,325	(a) 2.61:1 (b) 3.50:1 (c) 2.22:1	(a) 3.37:1 (b) 6.80:1 (c) 2.93:1
2.	*Tissue culture Dwarf-cavendish*					
	Traditional farming of local variety	Cultivation of *Dwarf Cavendish* Spacing: 2m×2m Seed rate: 10 suckers/cent	48,750	2,00,000	0.32:1	1.27:1
3.	*Nutrient management practices in coconut plantations*					
	Inappropriate application of nutrients and fertilizers	Nutrient management based on soil test data	15,000	50,000	0.75:1	2.00:1

able 15.3 : *(Contd..)*

Improved variety of ridgegourd, salad-cucumber and vegetable cowpea					
Traditional farming of local variety and Uneven spacing pattern and seed rate	Cultivation of improved variety and recommended spacing and seed rate (group farming)				
	(a) Ridgegourd: *Indam*-1222, 2m× 2m, 12g/cent	(a) 63,790	(a) 83,385	(a) 3.28:1	(a) 3.94:1
	(b) Salad cucumber: *Poinsetae*, 2× 1.5m, 3g/cent	(b) 44,225	(b) 62,725	(b) 2.55:1	(b) 3.32:1
	(c) Vegetable cowpea: *Arka garima*, 45× 30cm, 20g/cent	(c) 25,680	(c) 41,380	(c) 1.40:1	(c) 2.16:1

hitherto unutilised ponds for productive purpose thereby increasing the employment opportunities, income and food security. Integration of fisheries based interventions with livestock and agri-horticulture programmes enhanced the capacity of farmers to reduce the level of risk and uncertainties associated with overdependence on a single crop for their survival and sustenance. The coastal agro-eco-system analysis done as part of the Institute-Village-Linkage-Programme is replicable to other coastal agro-eco-systems with similar features. The technology assessment of various production systems in respective regions for further refined interventions could certainly improve yield scale in the long run. The results of IVLP clearly demonstrate its utility for preparing a comprehensive plan for the entire coastal agro-eco-system. Already there is spread effect and the fishery-based interventions are motivating more farmers to go in for diversified culture practices and bringing more area under aquaculture. The specific considerations to weaker sections of society and the women workforce in effect attract them also towards better production opportunities and promoting equity among the stakeholders. Thus the techno-interventions undertaken by IVLP based on a thorough understanding of the agro-eco-system not only implement the broad objectives of National Agricultural Research System (NARS) but also supplement and complement the livelihood options of farming community.

REFERENCES

Chambers, R. (1992). *Methods for analysis by farmers: The professional challenge.* Paper presented at the 12th Annual Symposium of the Association for FSR/E, Michigan State University, East Lansing, MI.

Chambers, R. (1992 a). *Rural appraisal: Rapid, relaxed and participatory.* IDS Discussion Paper 311. Brighton: IDS.

Conway, Gorden R., McCracken, Jennifer, A. and Pretty, Jules, N. (1987). *Training notes for Agro-eco-system analysis and Rapid Rural Appraisal.* International Institute for Environment and Development

NAAS (2001) Sustainable fisheries and aquaculture for nutritional security. Policy Paper 8. National Academy of Agricultural Sciences. New Delhi, India.

Pretty, J.N. (1995). *Regenerating agriculture: Policies and practice or sustainability and self-reliance.* London: Earthscan Publications; and Washington DC: National Academy Press.

Pretty, J.N., & Chambers, R. (1993). *Towards a learning paradigm: New professionalism and institutions for sustainable agriculture.* IDS Discussion Paper DP 334. Brighton: IDS.

Sathiadhas, R., Sheela Immanuel, Laxminarayana, A., Krishnan, L., Noble, D., Jayan, K.N. and Sindhu Sadanandan (2003) *Institution-Village-Linkage-Programme: Coastal Agro-Ecosystem and Interventions.* ISSN 0972-2351, Special publication No. 75, Central Marine Fisheries Research Institute, Indian Council of Agricultural Research, Cochin-14: 1-80.

Chapter - 16

Sayahnam — A Method for Timely Transfer of Technology

S. Ravi, and *D. Sajithbabu,*

Introduction

Agricultural production has evolved into a complex business requiring the accumulation and integration of knowledge and information from different diverse sources. The development of suitable extension system requires the combined efforts of specialists from many fields of agriculture and can be accomplished with the cooperation of farmers and policy-makers. Conventional agricultural research and development methods have not led to the creation of durable agricultural systems especially in the high-risk areas and have not been able to increase productivity substantially (James and Simon, 2002). In Kerala, extension service in the field of agriculture is mainly organized by the department of agriculture. Now, almost all the districts of Kerala is having Krishi Vigyan Kendra with a multi-disciplinary team can play an important role in imparting technical know-how to the farmers.

Krishi Vigyan Kendra, Sadanandapuram is in the field of agricultural extension and rural development since November 1994, imparting training to the farmers through on and off-campus trainings, making technological interventions in the farmers practice through demonstrations and making the farmers to assess and renovate technologies through trails on their own fields. A new programme named *sayahnam* was introduced by the Kendra to impart better understanding to the farmers about the common field problems and *in situ* remedies for their problems.

Location

Melila village was adopted by the Krishi Vigyan Kendra during 2001-02 and as many as three Front Line Demonstrations, three On-farm Trails were carried out with farmers' cooperation. It was found that some of the problems of the farmers were brought to the notice of the Kendra well in advanced stages, which at that time attained the level of point of no return or requiring high investment to contain the problem. The fact that earlier is better than late and prevention is better than cure was inculcated in the minds of the farmers and the ways to overcome the situation was discussed. It came to the nctice that farmers were ignoring some of the potentially dangerous problems at the initial stages due to ignorance and suffering a lot at the later stages. So it was decided to educate the farmers on their observations about the crops and field situations.

Modus Operandi

The panchayat ward member Smt. Anandhavally Amma had lent a helping hand in making the programme a success. It was agreed that a team of scientists from KVK would visit the village once in a week around 5.00 pm and the inauguration was held on 2nd November 2000. Thereafter it was a routine work of our KVK to visit Melila on all Saturdays in 2002-03. The time was fixed at 5.00 p.m. since almost all the farmers are free at 5.00 p.m. after their work. As it is the evening hour, the name of the programme is also called *sayahnam.*

Sayahnam

The programme started in November extended upto March; the attendance of the farmers thinned as the cropping season was over. The details of the *sayahnam* held at Melila panchayat is detailed below in Table 16.1.

Table 16.1 : Farmers participation details of *sayahnam*

Month	*No. of visit*	*Average participants*
November 2002	4	25.50
December 2002	3	18.67
January 2003	3	21.33
February 2003	3	12.00
March 2003	3	5.67
Total	16	17.19

Initially, the gathering was greeted by the KVK personnel and the discussion normally got started with the on-going farm activities. That was in the form of giving informal education to the farmers on the scientific nature of their activities and the technologies they are using. After that, the discussion centred on the problems they are facing in the field. Doubts and clarification sought by the farmers were discussed and a solution to that was normally derived from one among them. Each visit ended with field visit. This gave an opportunity to the scientist to explain what normally the farmers miss and also helped to educate all the farmers at one time.

Farmers were seemed to be worried much about the safety of their crop and wanted to know more and more about them. Most part of the time was spent on the discussion on the pest and disease problems. Existing technologies to the problems were discussed and the farmers were allowed to choose their solutions. If, there is a need for method demonstration, that was also arranged. It enabled the KVK to plan front line demonstrations and on-farm trails on a participatory mode.

Some of the drawbacks noticed during the programme were the loss of interest among farmers and only the farmers who were actually having the problems started attending the *sayahnam.* When the main crop was over by the end of February, there were a lot of dropouts.

Conclusion

Sayahnam is a way of contacting farmers, interacting with them, understanding their problems and delivering the possible technological solutions in a participatory mode. It also enables the extension workers to follow method demonstrations and make an instant field visit to clarify/educate the farmers wherever necessary. The cooperation of farmers and interest among them and the rapport of KVK with the farmers are the key factors responsible for the success of the *sayahnam*.

REFERENCE

James, T.J. and Simon, S.M. (2002). Grassroot innovations—Potentialities of

Chapter - 17

Impact of Extension Approaches
A Rajasthan Experience

Harphool Saharan

Introduction

Since a time long back in Indian history, agriculture has been the base for its economy. Today it is an anchor for achieving socio-economic growth as it accounts for 1/3 of the annual GDP and 2/3 of the national employment. The country has witnessed a spectacular growth in total foodgrain production from 51 million tonnes in 1950 to 211.3 million tonnes in 2002. It should be attributed to the hard work of the Indian farmers and to the sincere efforts of the extension workers in spreading the technologies developed by the research system.

Although, India has achieved food sufficiency, even then for its fastly growing population the household food and nutritional security along with environmental sustainability shall remain a concern of national importance in the times to come.

Since independence, organized efforts were made through various developmental and extension programmes to improve the socio-economic condition of the rural masses. Certainly the adoption of new and profitable agricultural technologies by the farmers have helped greatly to achieve the goal.

Keeping in view the national priorities various extension approaches at different points of time have been tried in India, but the ultimate aim of each approach was to accelerate the rate of adoption and dissemination of the technologies, so that the socio-economic condition of the clientele group can be improved. Some of the important extension approaches adopted in Indian extension includes:

1. Community Approach

Along with poverty and illiteracy the British rule inherited a lot of social problems like exploitation of farmers by the landowners, moneylenders and traders. In such a situation the efforts for overall development of the society on community basis were needed. The major developmental programmes initiated using this approach were Community Development Programme (1952) and National Extension Service (1953). Through these programmes many reforms in the areas of land ownership, co-operative credit were carried out and a strong base of infrastructure was created. But due to many reasons these programmes could not get desired success in increasing the food production and productivity.

2. Intensive Area Approach

Since independence, the country was passing though acute shortage of food and to solve this problem nothing significant could be achieved through Community Development Programme. In this situation the need for increasing agriculture production was urgently felt. Agriculture development was visualized as an instrument for rural development and intensive programmes for agricultural development such as Intensive Agricultural Area Programme, Intensive Agricultural District Programme, Intensive Cattle Development Programme, High Yielding Variety Programme etc. in selected areas were launched. As the result of 'green revolution' a historical success to increase the food production could be achieved. Very soon it was realized that being input (capital) intensive in nature, the benefits of these programmes have been harvested by some elites of the farming community, which resulted in the wider gap between the poor and the rich.

3. Target Group Approach

Extension programmes introduced during green revolution period were mainly technological development programmes while ensuring increase in production and productivity; these programmes failed to help the poor and neglected sections of the society.

Extension programmes introduced during green revolution period were mainly technological development programmes while ensuring increase in production and productivity; these programmes failed to help the poor and neglected sections of the society. Therefore, a new initiative was taken to improve the economic condition and correct the income disparities in rural areas through special programmes adopting the target group approach. Important programmes introduced in this period were Small Farmers Development Agency (SFDA), Marginal Farmers and Agricultural Labourers Programme (MFAL), Command Area Development programme (CAD), Training of Rural Youth for Self-Employment (TRYSEM) etc.

By all norms, these approaches paid rich dividend to the agriculture sector in terms of enhancement in quality and quantity of the products by a wider spread of modern technologies. But it was felt that

- Gradually the rate of increase in agricultural production slowed down.
- Many of the recommended technologies did not match with the farmers' need and resources. Hence a large group in the society remained non-adopter.
- Flow of the information was one way and the feedback was minimal.
- These approaches failed to make the farmer an active partner in the process of rural development.

In nineties, it was felt that people's participation in technology generation as well as in the planning and implementation of the extension programmes is of crucial importance. Therefore, the emphasis was laid on the participatory approaches, which resulted in the emergence of new concepts like watershed development, on-farm research and farming system research/extension, in the developmental and extension programmes. Recently, a new approach of "Integrated Technology Management" has been introduced by establishing Agricultural Technology Management Agency (ATMA) at district level through a World Bank funded project National Agricultural Technology Project (NATP).

ICAR started to work in the area of agriculture extension services in 1964 when the National Demonstration Programme was introduced. Further the services were extended through introducing extension programmes such as frontline demonstrations and lab to land. Later on, these programmes were merged in KVK. Presently, the KVKs are leading district level institutions in the field of agricultural extension with the revised mandate :

1. Conducting "On-farm Testing" for identifying technologies in terms of location specific sustainable land use systems.
2. Organise training to update the extension personnel with emerging advances in agricultural research on regular basis.
3. Organise short and long-term vocational training courses in agriculture and allied vocations for the farmers and rural youth with emphasis on "learning by doing" for higher production on farms and generating self-employment.
4. Organise frontline demonstrations on various crops to generate production data and feedback information.

As per the technical requirements of its implementation and objectives of the each extension programme require a specific extension approach or a suitable combination of the approaches. In the present chapter three major extension programmes of the KVK Sardar Shahr implemented using three different extension approaches have been discussed.

Frontline Demonstration

The main objective of Frontline Demonstration is to demonstrate newly released crop production and protection technologies and its management practices in the farmer's field under different agro-climatic regions and farming situations. Frontline Demonstrations are different from normal demonstrations in the sense that

- these are conducted in the close supervision of the scientists.
- only newly released technologies or those likely to be released in near future are selected.

- these are organized in a block of two to four hectares.
- training of the farmers is an essential component of the demonstrations.
- the scientists, to generate production data and feedback information study factors contributing to the higher production and field constraints of the production.

The extension approaches used in conducting the frontline demonstrations resembles to the intensive area approach used in IAAP and IADP for the reasons like :

- Complete package of the improved practices is demonstrated on the farmers' field.
- These demonstrations are conducted in an identified production potential cluster.
- Demonstrations are supported with the intensive extension activities such as field days, training of the beneficiaries and frequent visits by the scientists.

Since 1994 this KVK is laying out frontline demonstrations of oilseed and pulse crops in Churu district. The year-wise demonstration organized by this KVK has been given in Table 17.1.

All the demonstrations of Moth bean, Castor and Chick pea were conducted in rainfed conditions while demonstration of mustard and groundnut were conducted under irrigated conditions.

Training of Rural Youth for Self-Employment (TRYSEM)

Training of Rural Youth for Self-Employment (TRYSEM) was a facilitating component of Integrated Rural Development Programme (IRDP). These DRDA sponsored training programmes were conducted for the targeted groups of below poverty line rural youths who were interested in starting a common vocation for self-employment. During these long-term training programmes (each of 6 months) emphasis was given on acquiring the skills of the related trade. The village-wise detail of training programme conducted by this KVK under this scheme has been given in Table 17.2.

Table 17. 1 : Frontline demonstrations conducted by the KVK Sardar Shahr

Particulars		*No. of Demostrations*								*Total*
Crop	Variety	1994-95	1995-96	1996-97	1997-98	1998-99	1999-2000	2000-01	2001-02	
	Pusa Bold	10	–	5	–	–	–	–	–	15
Mustard	T-59	–	10	–	–	–	–	–	–	10
	Bio-902	–	–	–	6	7	10	–	10	33
	C-235	5	5	–	–	–	–	–	–	10
Gram	GNG-469	–	–	–	–	–	10	–	10	20
	GNG-663	–	–	–	–	5	–	–	–	5
Ground Nut	M-13	7	–	–	–	–	–	–	–	7
	Chitra	3	–	–	–	–	–	–	–	3
Moth Bean	RMO-40	–	10	10	10	10	10	5	10	65
Castor	G-4	6	–	–	–	–	–	–	–	6
Total		31	25	15	16	22	30	5	30	174

Table 17. 2 : Training programmes under TRYSEM arranged by the KVK Sardar Shahr

S.No.	*Name of Village*	*Number of Participants*	
		Male	*Female*
1.	Nainasar	–	18
2.	Ranasar Bikan	–	19
3.	Baniala	–	7
4.	Bharang	–	6
5.	Reri	–	16
6.	Bhurawas	–	6
7.	Gorisar	13	14
8.	Kanadwas	–	14
9.	Derajsar	5	–
10.	Sawai Bari	7	–
	Total	25	100

The approach used in conducting these training programmes can be said to be targeted group approach for the reasons like :

- These were conducted for the targeted groups of below poverty line rural youths.
- Subsidy to purchase the tool kits and establish their units was provided.

On-farm Trials

This type of trial is a set of procedures for adaptive research, which aims to develop recommendations for representative group of farmers. These trials are conducted using the participatory approaches as the overriding importance the farmers' perspective and participation is given at all the steps of the testing, viz. problem diagnosis, planning, experimentation and extrapolation. The objective of on-farm research is to identify existing inputs or practices that might help to solve major problems of many farmers in a defined study area.

Two on-farm testing titled (i) long intercalving period of cow and (ii) low milk production of cow were conducted by this KVK, the treatments/practices selected for the testing are given in Table 17.3.

Table 17.3 : On-farm testing the KVK Sardar Shahr

S. No.	*Village*	*Title of the Trial*	*Treatments Practices*	*No. of animals to those the treatment was administered*
1.	Bandhnou	Long intercalving period of cow	Calphos AD_3^+ + Panacur Bolus	8
			Cu-Co Tablet + Janova Capsule + Panacur Bolus	8
2.	Derajsar	Low milk production of cow	Farmer's practice + Mineral mixture	3
			Balance concentrate mixture+Mineral mixture	3

Impact of the Approaches

In the present paper an attempt has been made to measure the impact of the following three extension programmes of this KVK implemented using three different extension approaches.

Programme	*Approach*
(i) Frontline Demonstration	Intensive area approach
(ii) TRYSEM	Target group approach
(iii) On-farm Trial	Participatory approach

A sample of 112 respondents including 40 respondents of FLD, 22 respondents of OFT and 50 respondents of TRYSEM was selected. Although the nature of the technologies recommended for adoption to the clientele group was quite different, but the percentage of the adopter respondents was used as a common indicator to measure the effectiveness of the approach.

Impact of Frontline Demonstrations

Impact of the frontline demonstrations was measured in terms of the increase in the yield over local check and percentage of the respondents who adopted the demonstrated technologies /practices.

Data were collected from the forty respondents and results have been presented in Table 17.4.

Table 17.4: Adoption of the technologies demonstrated through FLD

S.No.	*Crop*	*N*	*Mean yield (qts. /ha)*		*Increase over control plot*	*Per cent*	*No. of adopters*	*Per cent*
			Demon-stration	*Control*				
1.	Moth Bean	20	3.977	1.65	2.33	141.03	16	80
2.	Chick Pea	10	8.41	5.97	2.44	40.87	8	80
3.	Mustard	10	12.34	8.45	3.89	46.03	6	60

The table shows that there was an average increment of 141.03, 40.87 and 46.02 per cent in the yield of demonstration plots as compared to the local checks, in the demonstrations of the crops of the Moth Bean, Chick Pea and Mustard respectively. Out of 20, 10 and 10 respondents selected for Moth Bean, Chick Pea and Mustard demonstrations. 16, 8 and 6 respondents were continuously using the recommended technologies thus in respect to the demonstrations on the crops mentioned earlier, adoption of the demonstrated technologies was as high as 80.00, 80.00 and 60.00 per cent. The overall adoption percentage was 75.00 per cent.

Reasons for Discontinuance/Non-adoption

It was found that non-availability of variety seeds and less resistant of varieties to insect, pest and diseases were the two most serious constraints in the adoption of the technologies given under FLDs and these were followed by the reasons like less market value of produce and frost injuries (it was observed in the mustard crops only).

Impact of TRYSEM

Table 17.5 reveals that the tie and dye vocation fetched maximum income to the respondents and the mean score of per annum additional income generated through this vocation was Rs. 3336. It was followed by the vocations of Dari weaving (Rs. 3024) and

Table 17.5: Adoption of the vocations learnt through TRYSEM

S.No.	*Vocation*	*N*	*Mean increased income (Rs./year)*	*No. of adopters*	*Per cent*
1.	Dari weaving	25	3024	7	28.00
2.	Tie and Dye	14	3336	7	50.00
3.	Stitching	11	2112	6	55.54
4.	Overall	50	–	20	40.00

stitching (Rs. 2112). The table also indicates that the maximum percentage of adoption (55.54%) was of the vocation of stitching followed by tie and dye (50.00%) and regarding *Dari* weaving it was only 28.00 per cent. The overall adoption of vocational learnt through TRYSEM was 40.00 per cent. The most serious constraint in adoption of the vocation was the problem of marketing of the product for the reasons (i) Less remunerative price of the product and (ii) non-availability of purchasers. Non-availability of raw material stood at second position in the sequence of seriousness of the constraints. It was followed by financial problems to start the vocation and lack of time.

Impact of On-Farm Trials

One treatment in each trial, which was found to be most suitable solution for the problem, was recommended to the farmers for adoption. It was observed that the treatment of 'Calphos AD_3^+ + Panacure bolus' to overcome the problem of 'long intercalving period of cow' was adopted by the 43.75 per cent of the respondents which reduced the intercalving period by 147.10 days and the treatment of 'Balance concentrate mixture + Mineral mixture was adopted by 66.67 per cent respondents which increased milk production by 1.79 litres/day/cow.

Conclusion

The results show that technologies given through frontline demonstrations were adopted by 80, 80, 60 per cent of the respondents in the crops of Moth Bean, Chick Pea and Mustard. While the recommendations emerged from on-farm trials were adopted by 43.75 per cent respondents in case of the treatment of 'Calphos AD_3^+ + Panacure bolus' to overcome the problem of 'long intercalving period of cow' 66.67 per cent in case of the treatment of 'Balance concentrate mixture + Mineral mixture' to overcome the problem of low milk production. Regarding TRYSEM, it was observed that overall the vocations were adopted by 40 per cent respondents.

It does not mean that the approaches by which we could not get desired maximum adoption percentage have lost their significance; rather these have established their importance in transforming rural India. Therefore, it can be suggested that suitable extension approaches should be adopted according to the priorities and objectives of the programme and need of the time.

REFERENCES

Choudhary, B.N. (1999). Krishi Vigyan Kendra (Agricultural Science Centre) : A Guide to KVK managers, Division of Agricultural Extension, ICAR, New Delhi.

Ray, G.L. (1991). Extension communication and management, Naya Prokash, 206 Bidhan Sarani, Calcatta.

Singh, Baldeo, Sharma, Poonam, Padaria, R.N. (2003). Technology Transfer and People's Participation: Core of Agricultural Development, Souvenir, Extension Strategies for Promoting Initiatives Among Farming Community: ISEE, Division of Agricultural Extension, IARI, New Delhi.

Chapter - 18

An Alternative Extension Strategy for Women Empowerment and Rural Livelihood Security

The KVK Kottayam Experience

K.J. Joseph., Pramod, and *Binu Johan Sam*

Introduction

This, indeed, is one of the success stories of Krishi Vigyan Kendra (KVK), Kottayam. A story on the germination and speedy growth to bearing maturity of a fruit tree by name 'Haritha', a women self help group, nurtured can do wonders in the field of women emancipation and enpowerment. A story that can be emulated and replicated anywhere in rural India.

The Genesis

KVK, Kottayam was established during March 2000 by Remandating Zonal Research Station (RARS, Kumarakom, Indian Council of Agricultural Research to take up additional functions of KVK with the existing infrastructure support from the ICAR under the National Agricultural Technology Project (NATP), with the major mandate of imparting need based vocational training programmes to farmers, farm women, rural youth and extension personnel of agriculture and line departments. Training need assessment of these groups were made through discussions and by conducting PRAs.

Rural women has been an important target group of KVK, Kottayam, right from its inception. So far 23 short-term trainings and two long-term vocational trainings were conducted exclusively for unemployed rural women. One such long-term training programmes was conducted for a period of six months from June to December 2001 on plant propagation techniques, bio-pesticide production and fruit preservation. On successful completion of the training, ten girls of the group registered themselves as a Self Help Group (SHG) under the Travancore Charitable Societies Act (1955) with the President of Kumarakom Grama Panchayat as Patron during February 2002. The SHG was christened as "HARITHA". The immediate objective was to start small scale commercial production unit of agricultural input like quality planting material, bio-pesticides, vermi compost etc. The long-term objective was to provide a viable income generating vocation to the unemployed rural women which will ultimately end up in escalation of their social status and improvement in the standard of living of their families.

The Setting

Utilising the services of HARITHA, KVK Kottayam established a bio-pesticide production unit and a horticultural nursery in September 2002. The raw material and infrastructural facilities were provided by the KVK and the work were carried out under the right supervision and guidance of KVK staff. A revolving fund was set up by transferring a small amount from the Zonal Research Station's as initial corpus. A Memorandum of Undertaking was signed between the SHG and the Kerala Agricultural University according to which the SHG would receive a fixed percentage of sale proceeds from the production centre as wages for their labour input. The rates were fixed at 20 per cent of the revenue from the sale of bio-pesticides and 30 per cent for planting materials. Bio-fungicides of *Pseudomonas fluorescens* and *Trichoderma viride, quality* planting materials like cashew and nutmeg grafts, ornamental foliage and flowering plants, lawn grass, medicinal plants and emerging crops like stevia and vanilla are produced here. These are being sold through the Agricultural Technology Information Centre (ATIC) of RARS, Kumarakom.

The Take off

Haritha was an instant success. The dedication and hard work of its members soon started paying dividends. The centre was 'air borne, within a few months. The details of production and sales of bio-pesticides are furnished in Table 18.1.

Table 18.1 : Production and Value of Bio-pesticides (month-wise)

Month and Year		*Production (kg)*		*Total Value*
		Pseudomonas	*Trichoderma*	
August	2002	115	156	13550.00
September	2002	0	189	9450.00
October	2002	101	0	5050.00
November	2002	1169	0	58450.00
December	2002	833	0	41650.00
January	2003	0	0	—
February	2003	90	398	24400.00
March	2003	669	0	33450.00
April	2003	0	0	—
May	2003	2631	0	131550.00
June	2003	2329	123	122600.00
July	2003	100	310	25500.00
August	2003	1300	0	65000.00
September	2003	2097	0	104850.00
October	2003	2767	638	170250.00
November	2003	2822	777	179950.00
December	2003	1980	193	108650.00
Total		19003	2784	1089350.00

Within a period of sixteen months from September 2002, around 21 tonnes of bio-pesticides were produced and distributed among farmers generating an income of Rs. 11 lakhs. Total turnover from the horticultural nursery is around Rs. 48,000.

Table 18.2 : Average per head earning of members (Month-wise)

Month & Year		*Ansamma A.S.*	*Elsamma K.V.*	*Linimol V.M*	*Manju M.K.*	*Premamol P.B.*	*Pushpa Kumari*	*Pushpavally K*	*Sheejamol A*	*Bank*	*Total deposit*
February	2003	1225	2275	2275	1558	1855	2188	2328	2170	2192	18066
March	2003	1705	2175	2725	1882	2295	2672	2823	2580	1437	20294
April	2003	345	360	525	368	450	525	547	495	0	3615
May	2003	2610	3002	3045	3219	3219	3306	3219	3219	1791	26630
June	2003	1152	1572	1612	1612	1252	2572	1612	1452	664	12500
July	2003	1152	1572	1512	1612	1352	1572	1612	1452	260	12096
August	2003	964	1804	964	1739	1880	1668	1857	1786	107	12769
September	2003	996	2202	2082	2240	2320	2254	2422	2500	1224	18240
October	2003	3059	4219	4155	4043	4384	4398	4495	4524	1123	34400
November	2003	2616	3260	3064	3296	3396	3096	3584	3464	254	26030
December	2003	3246	3790	3354	3348	3746	3634	4102	3945	155	29320
Average		1733	2385	2301	2265	2377	2444	2600	2508		
Total		19070	262323	25313	24917	26149	26885	28601	27887	9207	213960

The Impact

More than the income, the SHG activity has provided the members dignity and self-esteem, once they became co-bread winners of their families. The group is also engaged in social welfare activities like maintenance of village roads, general up keep of public health centre premises, financial support to people with chronic ailments in their locality, material and moral support to the inmates of orphanages, old age homes and other shelters for the destitute, assisting in adult literacy and various social awareness programmes in Kumarakom.

The Feedback

The members agreed that there had been a steep elevation in their social status after associating with KVK. "A job with a stable income has been my dream and HARITHA, to me, is like a dream comes true", says Miss Pushpavally, President of SHG. Miss Manju values most the new found social recognition which the stable income from KVK association has brought her. According to Miss Perumal, KVK has instilled in them confidence to plan and execute programmes successfully. Miss Elsamma and Miss Liny Mol are overwhelmed by the social acceptance and appreciation they have gained as earning members. Miss Ansamma is so happy that people both in and outside her family have started consulting her and seek her opinion. The KVK Association has enabled Miss Sheeja to support her family better.

Conclusion

Much of the unemployment in rural India is either disguised or wilful. To find out difficulties in opportunities is easy, but to find out opportunities in difficulties is not. This is where institutions like KVKs, can intervene. They can help rural people who are groping in the dark to locate those opportunities and motivate them to capitalize on these. KVK, Kottayam has three more registered self-help groups already working with it. Aiswarya, Lakshmi and Vembanad. Aiswarya SHG members, after completing a vocational training programme in fruit and vegetable preservation, is running a production centre for preserved and processed fruits. Lakshmi SHG is engaged in the commercial production of medicinal plants and mushroom spawn. Vembanad SHG members recently completed a training programme in fish farming and ornamental fish breeding. They have already started a production centre for commercial production of popular ornamental fishes. These SHGs are only in their infancy stage. We do not know how much successful they are going to be. But one thing is sure. Hard work will pay.

Chapter - 19

Alternative Extension Approaches

Agricultural Policies and Strategies for Profitability, Productivity, and Marketability

G. Perumal

Introduction

The first 60 years of the 20th century were marked by a sense of despair and frustration regarding our capability to achieve a balance between human numbers and the production of foodgrains and other agricultural commodities. In 1968, this mood of despair and diffidence gave wave to one of optimism and self-confidence in relation to our agricultural potential and our farmers' ability to adapt and adopt new technologies, a phenomenon which was as, social disintegration will be the result. The prosperity of the virtual world and the misery of the real world cannot co-exist for long.

Profitable agricultural industries producing high quality agricultural products in an ecologically sustainable manner that meets consumer needs, and based on a better understanding of the relationship between product prices and production costs, including environmental christened in that year as 'Green Revolution'. This agricultural transformation helped to strengthen our national sovereignty in foreign policy.

Agricultural progress will determine India's economic and political future. We can shape this future in a desirable direction through synergy among technology, public policy and farmers' cooperative action. If such a synergy can be achieved, India can become the foremost among the nations of the world in "farm power". If our agricultural progress is halted or reversed through inappropriate or inadequate agric policies and research priority costs. Achieving and maintaining sustainable agriculture has become one of the focal point. Agricultural Policies and Strategies are categorized under six divisions.

Agricultural Production

Agriculture, encompassing crop and animal husbandry, horticulture, forestry and agro-forestry, inland and marine fisheries and agro-processing, is the major determinant of the livelihood destiny of nearly 700 million people, out of 1 billion of India. The methodology of 'production by masses' characteristic of Indian agriculture cannot easily compete with products resulting from mass production technologies, until the power of scale both at the production and marketing ends is conferred on small scale production units, as has been done in the cooperative dairy sector. Therefore, trade policies, which impact on this destiny, are of vital concern to a majority of the rural population as well as to large numbers of the urban poor. There is need for an integrated trade strategy, which gives concurrent attention to home and external trade.

Development of Policies and Strategies

Existing models of agricultural intensification (that is, greater agricultural production on the same amount of land) assume that population growth, access to markets, and agricultural potential will lead farmers to adopt new technologies, intensify farming, and use resources sustainably. Yet despite policy reforms and structural and sectoral adjustment programmes designed to improve production and marketing incentives for farmers, low agricultural productivity, resource degradation, and poverty are still severe and worsening problem.

Policies and Strategies

Evolve and adopt best management practices for agricultural industries.

Facilitate agro-industry adoption of environmental management systems and other quality assurance processes to meet community demands and provide market advantage.

Promote training in and adoption of property management planning, including business and risk management.

Develop and adopt best practices for animal welfare on farms, in research and in transport and marketing sectors.

Cultivate and adopt alternative farming systems that allow adaptation to suit the variable soil and climate conditions.

Enhance (with appropriate safeguards) the diversity of genotypes used in agricultural systems to provide increased security and quality of production.

Explore and develop new domestic and export market opportunities for agricultural products (including organically grown products).

Produce agricultural products that satisfy market needs and meet industry and trade standards, where appropriate, inform the markets and adjust the standards to avoid unsustainable practices.

Amplify and adopt food production, processing and marketing strategies that improve human health.

Define and measure quantifiable indicators of the state of the agricultural and related environment, to allow better decision-making (both on-farm and regionally).

Promote the long-term economic, environmental and social benefits of sustainable agriculture to farmers and the general community.

Better inform rural communities of the role and appropriate use of agricultural and veterinary chemicals and biological pest control in sustainable agriculture.

Reduce the reliance of agro-industry on pesticides.

Diminish the risks caused by pesticides to the environment and to human health.

Elevate community awareness of the relationships between product prices and production.

Land Management

Agricultural land protected and managed for sustainable agricultural use with agricultural activity contributing to its protection and enhancement. In green revolution witnessed by our country, resulting from expansion in cultivated area, introduction of high yielding cultivars and improved management technologies has increased crop production. The increasing population has resulted the per capita availability of land, which is 0.14 ha. by 2000AD and further likely to decline to 0.08 ha. by 2020AD. Thus, the logical way to manage long-term soil fertility and productivity of soil with low cost inputs for larger agro based outputs may gain momentum.

Policies and Strategies

Develop awareness of the need to minimize adverse off-site impacts of land management practices.

Improve awareness and assessment of the causes, symptoms and impacts of soil degradation.

Encourage the protection, and promote the value of native vegetation and biodiversity in agricultural land management.

Amplify and adopt management practices and farming systems that conserve and enhance the health of soil resources.

Maintain effective barriers to the importation and spread of weeds, pests and diseases.

Promote and adopt integrated weed, pest and disease management strategies and best management practices (e.g. improved application technology, plant and animal breeding and biological control agents) that reduce the risks arising from use of agricultural and veterinary chemicals, with special recognition of the needs of farmers.

Use fertilizers to achieve maximum benefits in replacing nutrients used to produce agricultural products, with minimal adverse impact on the environment.

Monitor agricultural and veterinary chemical residues in agricultural products and in the environment; use the results to improve management performance.

Encourage and manage the agricultural use of organic wastes from urban, industrial and agricultural activities so that the components of waste are effectively used by plant growth or sustainably assimilated by the soil at the application site.

Water Use and Quality

Agriculture as an efficient and productive user of water contributing to the achievement of water quality and environmental flow objectives. Prime farmland is all the time going out of agriculture and groundwater depletion is proceeding at an alarming rate.

Our agriculture has reached a stage when proactive advice to farm families on land and water use planning based on an assessment of national and global market demand is vital for progress. At the national and state levels there is need for technical resource centres for monsoon management and water security. They should help to train rural climate managers who can help to maximize the benefits of good monsoons and minimize the impact of unfavourable monsoons.

Policies and Strategies

Develop, promote and adopt agricultural management practices that contribute beneficially to the management of rivers and riverside eco-systems, estuaries, groundwater and wetlands associated with agriculture.

Adopt efficient and effective water use practices (by irrigators, and other water users) that conserve and enhance the quality of the environment.

Evolve and implement land and water management plans in irrigated agricultural areas including adoption of best farming practice, irrigation system designs and decision support systems that overcome water degradation and provide for long-term irrigation community viability.

Provide government incentive programmes that improve the sustainable use of water by agriculture.

Participate collaboratively in the development of water reform policy by promoting and adopting river and ground water management plans.

Nature Conservation on Farms and Forests

Agriculture : making a beneficial contribution to the protection and management of the State's natural heritage and bio-diversity.

Nature Conservation on Farms and Forests

Agriculture : making a beneficial contribution to the protection and management of the State's natural heritage and bio-diversity.

Policies and Strategies

- Identify priorities for conservation on agricultural land.
- Adopt farming systems that conserve and enhance bio-diversity.
- Follow management plans that integrate nature conservation on-farm with agricultural production systems.
- Investigate and promote, using appropriate incentives, the agricultural and ecological benefits of conserving and supplementing remnant native vegetation.
- Recognise the integral role of trees, shrubs, perennial grasses and legumes in property and catchments management planning.
- Manage pest and diseases in ways that optimize agricultural and ecological benefits, and promote species conservation.
- Consider the establishment of sustainable farming of suitable native species.
- Ensure that the off-site effects of agriculture do not adversely affect areas set aside for nature conservation.

Rural Communities

Agriculture : making a beneficial contribution to Regional and State economics and contributing to the support of the State's rural people and communities.

Policies and Strategies

- Facilitate consultation among agricultural industries, indigenous stakeholders, government and the community on issues affecting rural policies and services.
- Improve the coordination and complementary nature of rural policies and services.

- Popularize the structural adjustment of agricultural industries and rural communities to enhance the sustainability of agriculture, including the provision and effective use of rural adjustment assistance.
- Provide opportunities for agricultural processing facilities to be located in rural areas and to strengthen their associated support infrastructure.
- Include the full range of agricultural industries in regional development strategies.

Integrated Management

Agricultural industries, communities and governments working together to achieve positive economic, environmental and social outcomes.

Policies and Strategies

- Implement policies and provide information to encourage the adoption of sustainable agricultural practices.
- Effect collaboration in the development, implementation and review of plans, policies and legislation relating to agriculture.
- Use Total Catchments Management and Land Care principles and plans to ensure that the off-site impacts of agriculture are considered and managed effectively.
- Confirm the equitable and efficient allocation of land other natural resources between agriculture and other sectors of the community.
- Guard land use planning is undertaken, where appropriate, in association with agriculture to avoid conflict that may jeopardize agriculture's sustainability.
- Ensure enactment of environmental impact assessment produces that result in the sustainable development of agriculture.
- Coordinate monitoring and reporting of the state of the agricultural and related environment, using measurable indicators that contribute to decision-making.

- Evolve and adopt agricultural activities and planning strategies that minimize impacts on community amenity from noise, dust and odour.
- Identify lands and farming methods best suited to specific agricultural industries and retain production options for those lands in the future.
- Develop and adopt appropriate planning mechanisms to avoid future conflict over land use (e.g. competing demands for land for agricultural, residential and recreational uses).
- Proceed efficient allocation and use of resources provided by the Central and State governments for agricultural resource management and improvement.

Sustainable Agriculture

In recent years, many people, both rural and urban become residents, have concerned that the agricultural practices used during the past 30 to 40 years may be damaging the environment and reducing food quality. Many people also worry that the economic strength and vigour of rural families and communities may decline unless farmers and communities can develop profitable farm and related non-farm activities. These concerns have prompted interest and discussion about sustainable agriculture.

A variety of labels have emerged over the past several years that attempt to classify sustainable agriculture. These include alternative, sustainable, low input, regenerative, bio-dynamic, and agro-ecological agriculture. Despite the different labels, they tend to focus on physical or biological relationships. Simply stated, sustainable agriculture focuses on farmers profit in the short-run, while preserving rural communities and natural resources in the long-run.

Some Ways to Achieve a More Sustainable Agriculture

- Identify, evaluate, and develop alternatives to those current farm practices that have the greatest potential to harm the environment or the health of farmers or consumers.

- Most people in agriculture are concerned about soil erosion, surface water contamination, groundwater pollution and depletion, food quality, and reductions in wildlife populations.
- Research, education and policy initiatives are needed for long-term solutions to these problems.
- Several practices that address environmental concerns already exist. Contour planting, minimal tillage systems, cover crops, crop rotation, and perennial grasses can reduce soil erosion and improve profitability.
- Many sustainable agriculture practices substitute on-farm resources and management for purchased inputs. Increasingly, some farmers use crop rotation systems, animal manures, and mechanical cultivation to maintain soil fertility and productivity. Develop profitability strategies that focus on diversified farming systems that effectively use both external and internal inputs assist farmers creatively marketing their goods.
- If adequate management and labour resources exist, diversification reduces financial risk.

Diversification hedges against drought and economic pressures from increased input costs, commodity price declines, and regulations that affect the supply of certain commodities. Diversified farm strategies may also include premium-prices products, such as organic or superior quality products. In other words, the focus may be on higher quality products that are more valuable as opposed to focusing on simply increasing production. A diversified strategy also seeks to access local, regional, national, and international markets.

Develop policies that maintain a diversified farm structure and rejuvenate rural communities. This is a "big picture" perspective of sustainable agriculture. A diversified agricultural structure includes part-time farmers, full-time farmers, single-family farmers, and larger multi-family partnerships or cooperatives. A diversified rural community structure also focuses on creating and maintaining non-farm enterprises that complement farm families' rhythms and needs for off-farm employment featuring well-paying jobs and adequate benefits. The idea is to build both income and an "infrastructure" of business and community services, (Steve Stevenson, 2003).

Satellite Farm Approach for Sustainable Agriculture

(Tamil Nadu Agricultural University, Coimbatore, India)

Satellite Farm Approach: For sustainable Farm Enterprises and Human Development

The Satellite Farm Approach is the outcome of action research, and its unique feature is that farmers become the focal point of all developmental efforts facilitated by field level extension workers and scientists. *In a nut shell, a Satellite Farm is of any size, diversified, and developed over a period of time by applying appropriate technologies generated by the research system and recommended by development departments, NGO, media and other supporting agencies.*

Satellite Farm Approach is a bottom-up approach that starts by trying to understand the dynamics of particular situation—neighbouring farmers, their daily tasks, needs, and aspirations, as well as constraints hindering their acceptance of different enterprises for progress. This approach is a farm family centred process that involves a holistic analysis of a given farm and family situation. It also demonstrates the application of two basic principles—community participation and sharing interdisciplinary work methods (G. Perumal and K.A. Ponnuswamy, 2003).

Advantages

- Facilitates desirable 'farmer-farmer-extension worker-scientist' interaction.
- Provide greater scope to extension workers and farmers for experimental learning.
- Gives opportunities to study and understand interdisciplinary and inter-enterprise technology, behaviour and also the complementary flow of bio-resources within the farm.

- Enables identification and understanding of the use of Indigenous Technical Knowledge.
- Helps extension workers and scientists to understand the ability of farmers to manage crisis situations based on their native wisdom and experience.
- Helps farmers develop their own skills in choosing the appropriate combination of enterprises.
- Creates a cooperative attitude among farmers for better production and marketing of various agricultural products.
- Does not warrant any additional cost of manpower on the part of the development departments.
- Above all, it ensures cost-effective and quality extension work.

Food, Farming and the Future: Field Realities, Needed Reforms And Interventions

There is growing consumer demand for high quality food, produced safely in ways that do not harm the environment, yet even though consumers want high standards, they still want to buy food as cheaply as possible. In these circumstances a tendency has developed for realities to try to pass on higher production costs, not to the consumer who demands them, but to the producer. As costs are passed on to the producers, wherever they live and work in the world, rural communities still further, and impoverished, widening the gap between urban and rural communities still further, and placing additional pressures on rural communities and therefore, ultimately, the environment.

The conference of Common Agricultural Policy, was set against the background of the need, to further reform—both in its regional structure and 1992 reformation has kept farming profitable (Farmer's Link-UK, 2003). However, many people all over the globe believe that further change is inevitable (and indeed desirable) for the benefit of rural societies. Farmer's Link continue to facilitate debate on the future of food and farming by offering a regional forum where farmers, consumers and policy-makers can discuss what they wish to see happen in the future to food and farming, as well as the sustainability of current agriculture and trade policies.

practices lead to higher food prices for consumers and wasted tax money. Subsidies and price controls can reduce the incentive to innovate and increase efficiency. But changing these practices is a politically sensitive issue and requires thorough analysis of policy options (Suresh Babu, International Food Policy Research Institute, 2003).

In improving Agricultural Policies in India, Subba Rao, (2003). stressed Policy-makers need to use research information on the impact of current practices and on the potential effect of reforms.

The dire need of the hour for the Nation in terms of agricultural reform and modalities are to be decided by the policy-makers, academicians, farmers, and representatives of non-governmental organizations and the private sector, based on their recent experiences.

Less Intensive Farming and Environment

Integrated pest management, integrated farming systems, integrated production and outline what is happening in practice. By David Buffin

Less intensive farming (LIFE) demand a holistic pattern of land use, which integrates natural regulation processes. This helps to reduce the need for off-farm inputs such as, fertilizers, pesticides etc. Crop rotation, cultivation and soil management, cultivars resistant to pests and diseases, sowing date, nutrition, bio-diversity etc. are optimized, they all form important components of integrated farming systems.

Profitability can be maintained with less-intensive production (Vic Jordan, 2003). Results from arable production have shown that herbicide use has been reduced by 19 per cent, fungicide use by 84 per cent and insecticide use by 100 per cent. The management of input has not yet been optimised.

In relation to pesticides, the following practices for LIFE farming have been adopted :

The use of chemicals that are more selective;

The use of minimal dosage and frequency of application.

Spot treatment or inter-row applications (not yet implement on pilot farms);
Low volume sprays and more careful timing of application.

The following strategies are recommended in order to reduce inputs :

Modification of cropping sequences to increase crop diversity;
Use of tillage systems that favour natural control key pests (pests, diseases and weeds) improve soil structure and reduce demand for external nitrogen;
Development and use of pest thresholds with decision models and non-chemical methods to reduce agro-chemical input-use of biofertilisers;
Modification of field margins to encourage natural pests.

Indian Scenario

Sukhdev Singh Dhindsa Hon'ble the then Minister of Chemicals and Fertilizers, Government of India (2003) said that the National Average Consumption of Fertilizers in India, in terms of NPK has shown a steady growth during the last decade and is now estimated to be in the range of 90-95 kg per hectare for the last two to three years (2000-2003). Unfortunately, indiscriminate use of fertilizers without scientific basis has had an adverse impact on soil fertility. There is a need to promote integrated nutrient management system that envisages simultaneous use of chemical fertilizers, organics, and bio-fertilizers.

Value Added Products

Value added products are made from other primary agricultural products. Some are turning waste straw into board or using it to build stronger auto parts. Others are making nutritious animal feed from waste eggs. Producing value-added products is a growing trend that benefits both farmers and economy. Farmers can get more financial return from their agricultural products, which is important in this day of the farm crisis when many are struggling to keep afloat.

The food processing Industry sector in India is one of the largest in terms of production, consumption, export and growth prospects. The government has accorded it a high piority, with a number of fiscal reliefs and incentives, to encourage commercialization and value addition to agricultural produce; for minimizing pre-post harvest wastage, generating employment and export growth.

Important sub-sectors in food processing industries are: Fruit and Vegetable Processing, Fish Processing, Milk Processing, Meat and Poultry Processing, Packaged/Convenience Foods, Alcoholic Beverages and Soft Drinks and Grain Processing etc.

Organic Farming

Agriculture and Health Promoting: A Shift to Natural Food

The natural farming promotes and provides the nation with healthy food and other raw materials produced in a sustainable, humane and pollution-free way. Thus the high quality and healthy food thereby helps in reducing disease and medical costs. In turn it may be useful to increase and diversify rural employment, thereby bringing balance and vitality to country communities. Long-term, our policies will lead to a diversified, ecologically sustainable, aesthetically pleasing countryside. We will not allow short-term gain to compromise the future health and well-being of the population and the ecological integrity of the environment.

Organic Crops can be as, or more Profitable as Conventional

Organic cropping systems in the Midwestern United States can be as profitable, or even more profitable, than most conventional rotations (Henry A. Wallance, 2003). To help farmers understand the profitability of organic agriculture, the Wallance Institute report, by policy analyst Rick Welsh, analyzes studies comparing organic and conventional grain cropping systems. The report also reviews past and current research on the conditions under which growing organic crops is profitable. There has been dramatic world-wide growth in the production of and demand for, organically produced agricultural products. In addition, consumers have consistently been willing to pay premium prices for organic products, which have often caused processors to pay premiums to farmers for organic grains. However, the study found that premiums are not always necessary for organic systems to out-perform conventional systems.

When the organic systems were more profitable, it was due to one or more factors, including:

(i) The organic system having lower production costs.
(ii) The net returns for the types of crops in the organic rotation were higher than the net returns for the types of crops in the conventional rotation.
(iii) Organic systems are drought hardy and can outperform conventional systems in drier areas or during drier periods.

Marketing Aspects of Organic Agriculture

State departments of agriculture, and other interested parties that view organic agriculture as a potentially fruitful area for environmental management and rural development, might invest resources to develop marketing information useful to organic farmers. In addition, extension services could assist organic farmers in developing or locating market outlets or developing individual or locating market outlets or developing individual or cooperative marketing strategies.

In India, self help groups, non-governmental organizations, voluntary associations may be involved for promoting organic agriculture programmes. Many Agricultural Universities started issuing certificates for the standard-quality organics.

Interventions

The global necessity to increase agricultural production from a steadily decreasing and degrading land resource base places considerable strain on the fragile eco-system. Land must be cultivated more intensely, water and labour be utilized more optimally and fertilizer and other purchased inputs used more efficiently. While the use of mineral fertilizers is the quickest and surest way of boosting crop production, their cost and other constraints frequently, deter farmers from using them in recommended quantities and in balanced proportions.

Organic residues both of plant and animal origin are by-products of agricultural activities and thus virtually a source of plant nutrients. Organic manure, though bulky in nature and with low nutrient content, have the property of improving soil's physical and microbial conditions, and thus enhance fertilizer use efficiency when applied in conjunction with mineral fertilizers. Organically produced products a fetch premium prices.

Integrated nutrient management differs from conventional nutrient management in that it is more explicitly considers nutrients from different sources, notably organic materials, bio-fertilisers, nutrients carried over from previous cropping seasons, the dynamics and transformation of nutrients in soil, interaction between nutrients, and the availability of nutrient in space and time in relation to the nutrient demand by the crop. Integrated nutrient management is to be considered an integral part of any sustainable agricultural system for profitable Agriculture.

Demographic trends in India have two important implications for agricultural research and development. First, more than 50 per cent of the over 1 billion population belong to the age group 21 and below. Unless farming becomes both intellectually stimulating through the pathway of profit-based precision farming, and economically rewarding through value addition to primary produce, it will be difficult to attract or retain youth in farming.

The other demographic trend is increasing urbanization. Soon, 50 per cent of the population will be living in towns and cities. Urban agriculture and urban green belts offer opportunities for jobs and income, as well as for improving the urban environment and quality of life. Schools and colleges in urban areas can promote with the help of agricultural universities and institutions and green belt development, which can help to promote symbiotic links between rural and farmers and urban consumers. Also they will help to generate more non-farm jobs, which is an urgent need in rural areas (Swaminathan, 2003).

There is need to restructure research strategies in a manner that strategic, anticipatory and participatory (i.e. with farm families) research all receives adequate attention. Similarly, extension services should become farmer owned and controlled and should become capable of converting generic into location specific knowledge essential for taking to precision farming methods.

Technology research is needed on intensification of intercropping and mixed cropping techniques that increase output, incorporate cash perennials, and increase crop density while protecting the soil.

The Satellite Farm Approach is the dire need of the hour as it is holistic, induces farmers to develop their own skills in choosing the appropriate combination of enterprises, creates co-operative attitude for better production and marketing of agricultural products.

At the National and State levels there is need for Technical Resource Centres for monsoon management and water security. They should help to train Rural Climate Managers who can help to maximize the benefits of good monsoons and minimize the impact of unfavourable monsoons.

Import policies relating to farm commodities should be based on a careful assessment of their impact on all those who depend upon agriculture for their livelihood security. Government should bring out a White Paper on the World Trade Agreement and Indian Agriculture.

The rural knowledge Centres should provide computer aided and Internet connected information services, so that farm families have timely and relevant meteorological, management and marketing information of agricultural/agricultural industry's end produces.

REFERENCES

David Buffin (2003) Pan International Website, Natural Collection, UK.

Henry A. Wallace (2003) Organic crops can be as or more profitable. *Alternative Agricultural News,* 14, 1. USA.

Perumal G. and K.A. Ponnuswamy (1997), India 2, Extension, Home Site Contents Search World Bank, Website.

Steve Stevenson (2003) Research Brief—Sustainable Agriculture: Balancing Profits, Human and Natural Resources, Centre for Integrated Agricultural System, University of Wisconsin, USA.

Subba Rao, I.V. (2003) Improving Agricultural Policies in India, Workshop on 'Agricultural Policies for the 21st Century, From Vision to Action' India.

Sukhdev Singh Dhindsa (2003) Union Minister Chemicals and Fertilisers, Government of India, Home Website.

Suresh Babu (2003) International Food Policies in India, Website.

Swaminathan, M.S. (2003), 'Shaping Our Agriculture Future', Indian Science Congress, New Delhi.

Chapter - 20

The New Zeal of Extension Approches Adopted by KVK, Gandhigram

P. Venkatesan and *S.K. Gopal*

Faith and doubt both are needed not as antagonists, but working side by side to take us around the unknown curve. The job of extension personnel is not the bed of roses alone but it is also a bed of thorns. The researchers neither have the time nor are they equipped for the job of persuading the villagers to adopt scientific method and to ascertain from them the rural problems. Similarly, it is difficult for all the farmers to visit the research station and obtain first hand information. Thus there is need for an agency to interpret the findings of research to the farmers and to carry the problems of the farmers to research to solution. This gap is filled by the extension agency.

Ideas are like seeds. They also require proper season, time and place to take root and grow. But more importantly, they also need the right technology and infrastructure to support for achieving the full potential. For long we have been talking of placing the farmer at the centre stage of all research and extension efforts and developing models of participatory technology generation and dissemination. We believe the time is now ripe for us to build on our past experiences and fashion such a system.

The infrastructure, manpower and resources are present. Economic liberalization has changed our orientation from a pure public delivery system to a more facilitative role in channeling the increasing capabilities of the private corporate section and NGOs. Our experience has taught us that sustainable agriculture requires no mere consultation but total involvement and participation of farming community for which getting farmers organised to manage their own natural and other resources and economic enterprises is the key. Research and development efforts have brought us on the threshold of a quantum jump in the availability of a variety of crop-production technology generation, information processing technologies, participatory extension and communication and value addition technologies.

"Nation prospers as the village prospers" said the Father of the Nation. Gandhigram, a village bloomed in 1947 by the direct Dhahran and blessings of Mahatma Gandhi is committed to rural development. Gandhigram Rural Institute (GRI), over a period of two decades of educational experiment in rural higher education attained the status of Deemed University in 1976. The 'Krishi Vigyan Kendra - KVK (Farm Science Centre), of GRI yet another effort in this direction was established in July 1989 specifically to cater to the needs of the peasants of the Dindigul District of Tamil Nadu funded by the Indian Council of Agricultural Research (ICAR), New Delhi. The KVK has made its mandate in its letter and spirit.

1. Organising vocational training programme in agriculture and allied enterprises.
2. In-service training of field level extension officials.
3. Front Line Demonstration on oilseeds, pulses, cereals and horticultural crops to reveal the potential of new technologies.
4. On Farm Testing crop production, horticulture, livestock production etc.

Unique feature of KVK lies in the core centre of agricultural extension system, comprising scientists from different disciplines of agriculture. A farmer who approaches KVK can get information in all relevant areas of his farming. The other important feature is that the KVK provides intensive training to the farming community through the 'On' and 'Off' the campus programmes.

Technology Information Spread — Farmers Discussion Groups (FDG)

> "To keep the farmers well informed of the programmes and for quick dissemination of technologies, Farmers Discussion Groups have been organised at the village level. About 100 such groups are functioning and are encouraged to meet periodically to sort out and debate on the burning issues. Similar groups are organised involving farm women to motivate the habit of saving on the lines of 'Self Help Groups' for women employment."

Nexus between Farmers and Experts

Training programmes are offered by KVK in the fields of agriculture, horticulture, agro-forestry, animal husbandry, home science and extension education. The programmes are planned in advance of the ensuing agricultural season taking into consideration the existing cropping pattern. The suggestions of the conveners and members of Farmers Discussion Groups and Development Departments are given due importance. Recently the use of Participatory Rural Appraisal Techniques at various stages of programme implementation is being followed, not only to enhance people's participation; but also to reach the ultimate objective with ease.

KVK, GRI has been a propeller in the development of villages by imparting training to farmers. Various training programmes to farmers as per their felt needs are being tailored and imparted. It is equally important to find out the training methodology, which is mostly preferred.

Training needs of each subject matter area was asserted over a three-point continuum such as most needed, needed and not needed and they were quantified by assigning 3, 2 and 1 scores respectively. Index was computed for different items for each subject matter areas. The obtained score for each respondents was worked out by multiplying the number of respondents with their corresponding scores given (i.e. most needed 3, needed 2 and not needed 1) the obtained scores of all the items were added and divided by the potential score to arrive at Training Need Index (subject matter).

$$\text{Training Need Index} = \frac{\text{Summation of obtained score}}{\text{Potential score}}$$

Obtained score of an item = No. of respondents × score given to the corresponding subject matter

Total no.of respondents × highest rank given for subject matter

Subject matter-wise training needs were assessed and processed accordingly.

Another method of protruding into the thrust area is by using participatory rural appraisal. There are different types of participation in information giving, participation by consultation, participation for material incentive, functional participation and interactive participation.

Diagramming and visual construction by local people gives them a share in the creation and analysis of knowledge, provide a focus for dialogue, which can be subsequently modified and intended. Local categories, criteria and symbols are used during diagramming. After usage of all the maximum possible tools of PRA, SWOT analysis is done to identify the felt training needs of the farmers of a particular village. By using the PRA tools and training need index methodologies the training needs of the farmers are assessed.

Next step in training is to build the training competency of the extension personnel who has to acquire, understand and apply technical agent to correctly handle relevant technical materials and equipments *Technical Competency*. The extension personnel has to understand the market forces to advise and guide the client system for maximization of benefits. It includes the ability of extension agent to organise economic activity with the client system such as formation of beneficiaries committee, cooperative society, arrange bank credit etc. *Economic Competency. The Scientific Competency* of the extension wing is the belief that science can change human destiny. To conduct simple trials and demonstrations in the clients own situation and interpret the result to them is the occupational competency. Like-wise *Communication Competency* and *Social Competency* of the extension wing is built up. The training materials visual, audio-visual, projected and non-projected visual aids are selected for the farmers based upon their training needs. Internal structure and policies for the implementation of the training programmes for the farmers are formulated. Action plan are prepared on monthly, quarterly, half yearly and annual basis for the implementation of the training programmes.

Training programmes or the seminar or the campaign is implemented with functional linkage, diffused linkage, collegial linkage, and normative linkage and enabling linkage. After the implementation reconsideration of the programme is done. Reconsideration is done in each steps of training, i.e. while planning, executing, developing and implementing stages.

Second Mandate of KVK is to conduct "On Farm Testing (OFT)" for identifying technologies in terms of location specific sustainable land use system.

On farm testing is conducted in the farmer's field on such problem where the appropriate technologies are not available for particular agro-climatic situation to transfer the relevant search information available does not suit the situation from the point of view of the farmers.

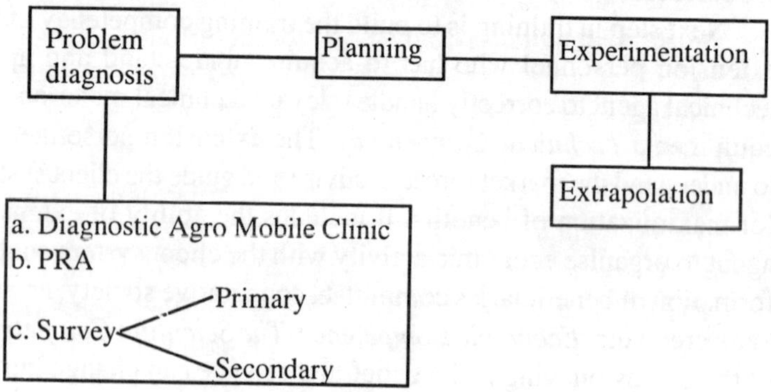

The technologies generated at the research station may not provide expected results at the farm level. To refine the generated technologies to suit to the local situations, OFT is carried out with farmer's participations. Diagnostic Agro Mobile Clinic is conducted in the KVK with adopted villages and farmers problems are diagnosed and documented. Participatory Rural Appraisal tools are also used in problem diagnosis. Then triangulation is made with the survey to know the felt problems faced by the farmers. Next step is to avert the problem, for which planning is done with the subject-matter specialist, the area is demarcated and the farmer is identified and action plan is modulated on formulated for the programme is implemented in experimental basis with two splits/ two treatments. Then the extrapolation of the work is done. Monitoring work is done in each stage of execution. The third mandate of KVK is to organise "Front Line Demonstration on various crops to generate production data and feedback information".

The field demonstrations are conducted under the close supervision of the scientists of National Agricultural Research System (NARS) are called Front Line Demonstration, because the technologies are demonstrated for the first time by the scientists themselves, before being passed on to the main extension system of the State Departments of Agriculture. The objective is to demonstrate newly released crop production and protection technologies and its management's practices in the farmer's field under different agro-climatic region and farming situation.

While demonstrating the technologies in the farmer's fields, the scientists are required to study the factors contributing to higher crop production, constraints of production, and thereby generate production data with feedback information. FLDs are conducted in a block with two to four hectares land in order to have better impact of the demonstrated technologies on the farmers and field level extension functionaries.

The fourth mandate is to organise training to update the extension personnel with emerging advances in the agricultural research on regular basis at present the KVK trains the extension personnel of Government Organisation, Non-Government Organisation and other personnel of different agencies. This KVK also follows a paradigm for implementation of the same.

1. Pre-training: Expectation and Motivation

To learn, people need to feel the need for learning. This is true for any kind of learning. The need can be desperate, as for survival or for even economic survival. More often, learners need more skill to qualify for a particular job or to feel more effective in it, to understand particular things or events, or to feel that they are continuing to grow

Some people feel that they are continuing to grow. Some people feel this latter need readily and take pride in their ability to learn new habits. Other reject on play it down, because learning for them as akin to an admission that they are not already competent. These two types of people and other faces the prospects of training with very different expectation and are right in thinking that they will device very different experiences from same training sessions or courses. Here, the essential point to group in that participant's motivations and viewpoints will determine their focus of attention and learning.

2. Training: Exposure to Opportunities for Learning

Expectations, whatever they may be, are brought by participants to a training programme. Exposure to the learning opportunities training offers will hopefully result in a series of five stages. Each series commences with the selection of an item to learn and includes with a new behaviour becoming habitual and participants may undergo several times the same impact and may well interrelate them.

(a) Selection of Some Items for Learning

Each participant, and nor the trainer, "decides" the focus of his/her attention and his/her expectations from it. "Decides" has been placed in quotes to signal that many factors, not all conscious or apparent, go into a participant's choice of focus.

(b) Initial Trial of Experience

The participant now explores and uses the items selected. He thinks about it, and translates his/her interest into specific behaviour. He tries it out. Does it fit into this way of thinking? or that way? into other people's expectation? his/her purpose in this initial trial is to see its effect on the organized system of thinking, including her own self-image.

(c) Feedback from the Initial Trial

The participant's experimentation may be rewarding or discouraging, or merely open-ended. This depends on his/her own reaction to the trial, and the reactions of other participants and the trainers.

(d) Reinforcement and Continued Practice

Satisfaction from positive feedback reinforces the participant's new pattern of behaviour. This amounts to practice. Repeated practice followed by repeated satisfaction will result in establishing the new pattern of behaviour firmly.

(e) Internalizing What She has Learned

Learning bases on repetition tends to be mechanical and barren until the participant ceases to be self-conscious of it and uses it routinely as a part of her habitual pattern of behaviour. This process of assimilation takes time. The new item, are the modification of an established habit, effects many parts of the participant as a person. It becomes woven into her experiential texture and makes her a somewhat different person.

3. Post Training Phase

The training session is the programme ends. The participants leave for home with whatever they have learned. If things have gone well, they go back to their organizations with motivation and enthusiasm newly heightened by the satisfaction of learning, they are eager to put what they have learned to use on the job. In doing so they behave differently, in fact from the way they behaved prior to training. Now, back on the job, they have to add more dimensions to their learning, an organizational one.

(a) Organizational Supports are Lack of It

Once again the participant watches for feedback. How do others react to her new behaviour? If her colleagues and superiors approve of it, she is encouraged and inclined to continue to behave in the more effective way. If they ignore they return or any indication of change in him, behaving as though he/she had never been trained, he/she will be discouraged from persisting with attempts to use what she has learned. Some anxiety will persist until the reception is reasonably clear.

(b) Improvement on the Job

If the final analysis, training is successful if participant's job effectiveness improves are expected. Verbal support for more effective behaviour does not suffice to ensure this result. Other pre-conditions, such as adequate equipments, basic attitudes of colleagues, and organizational policy, have to line up with the change.

Discrepancies on the job lead to participant frustration at the very time that their expectations are highest. This final failure of training is all to well known to young people returning from training overseas. Others fear it to such an extent that they do not return to their home even to try. Alternatively, when conditions are favourable, changes actually take place. More important for the long-term is the favourable outlook such success creates for further learning in the work situation itself. The participant's learning depends on continuity. It must suffer no breaks during any of the several stages. The above extension models are formulated by this Kendra to avoid or rectify the hindrance faced in the implementation of the programmes in the early efforts.

Chapter - 21

Krishi Vigyan Mandal New Approach of Farmer to Extension

S. Sonune, K. Dattatri and *M. Hegde*

In India, like in United Kingdom top down approach was followed in the transfer of technologies to the farmers in which the technologies to be transferred are decided by the bureaucrats and technocrats and the lowering extension workers job is to simply carry it to the farmers. The success or failure of the implementation will be assessed based on the increase or decrease in the yield of targeted crops. This method found to be not only expensive but also non-responsive to the needs and aspirations of the farming community.

Of late in India like in many countries agricultural extension is based on participatory approach and farming systems development approach has gained acceptance readily. Here the assumptions are that the ministry and its research system may not have the answers to farmers questions, in fact, may not even be aware of some of farmers' questions. Therefore, extension staff and research staff team collaborate with each other and also with farmers and they set out to both generates appropriate technology and transfers it to farmers wherever it fits. Unlike in earlier time, wherein, bureaucrat and technocrats used to be the decision-makers, where as now farmers are the decision-maker and others are facilitators. In our type of democratic governance it is very much essential that for any programme to be successfully implemented policy-makers support is very much essential.

In a country like India wherein agriculture is practiced in diversified situations encompassing various agro-ecological situations no single system of technology transfer can be highly successful. Hence, when we review the effectiveness of various models used during last 50 years it is seen that each one has succeeded in pockets. It is also essential to make changes in our approach keeping in view the changing circumstances.

Now-a-days group methods are increasingly used in agricultural research extension and other development activities (Kumar, 1987). Group meetings can help in eliciting the farmers' ideas and problems, in discovery and enhancing their knowledge and innovations and in trials and technology development. Many of the agricultural activity like mechanization and pest management can be successful when only group approach is followed and all the farmers of the area practice it. It is also to be noted that by group approach reduction in cost of production can be achieved and benefits are shared. Once the intact groups are formed it becomes easy to use them for implementing other programmes.

The world-wide experience of different farmer groups have shown that farmer groups can play three types of roles, viz. representation role, technology role and service role (Arnaiz 1995).

Farmer groups' refers to groups composed mainly of members of the rural community, along with one or more agricultural researchers and/or extensionists. Purposes of farmer group can include:

(i) Building interaction and communication between researchers and farmers, eliciting and exchanging information from farmer to farmer, from farmer to researcher and from researcher to farmer.

(ii) Analysis by farmers with research support of their problems and needs, reinforcing and fostering their own knowledge and capability.

(iii) R & D with the choice, design, conduct, monitoring and evaluation of experiments.

(iv) Extension from farmer-to-farmer and the diffusion of innovations.

(v) Empowerment of farmers to organize for action or to share the resource.

problems and needs, reinforcing and fostering their own knowledge and capability.

(iii) R & D with the choice, design, conduct, monitoring and evaluation of experiments.

(iv) Extension from farmer-to-farmer and the diffusion of innovations.

(v) Empowerment of farmers to organize for action or to share the resource.

Krishi Vigyan Mandal – A Model

In Jalna district of Maharashtra, India innovative farmers together formed a group called "Krishi Vigyan Mandal" to exchange ideas, latest technical information and innovative methods that can be adopted them in their farms. The main purpose of the farmers in formation of mandal is to get quick information and diffusion of information in large number of group through farmer-to-farmer extension. The mandal was established on 29th September 1997 in the pemises of Krishi Vigyan Kendra, Jalna. The idea of forming this mandal was crystallized when the farmers came together at KVK Jalna to attend a function.

Jalna district of Maharashtra falls under Central Maharashtra Plateau agro-climatic zone having sub-tropical climate. The annual rainfall of the district ranges from 650-700 mm with an average of 725.8 mm bulk of the rainfall received during southwest monsoon. The soils are medium deep with low rainfall irrigation potential is 10 per cent to total cropped area and the major source of irrigation is wells. From the above it can be said that the district falls under Complex Diverse and Risk prone (CDR) system of agriculture.

Objectives : According to the group the objectives of the Mandal are as follows :

- To exchange ideas and thoughts among the group,
- To know and adopt latest technical know-how from scientists and resource persons,
- Transfer of technology amongst farming community through the members of Krishi Vigyan Mandal,
- To form specialized interest/commodity groups for specific

is a Farm Science Centre funded by the Indian Council of Agricultural Research (ICAR) and works under the administrative control of an NGO, i.e. Marathwada Shethi Sahaya Mandal. KVM is totally an informal group of progressive and innovative farmers of Jalna district and few from neighbouring districts. Initially the group consists of about 314 members and the number now increased up to 1500. The members of the group voluntarily contributed Rs. 100 as annual membership fees for meeting out miscellaneous expenses.

Network of KVM

The group consists of key, nodal and follower farmers. Key farmers are initially joined in KVM and are responsible for successful dissemination of information to other farmers; nodal farmers are attached to KVM through key farmers and disseminating the information to other farmers and also in touch of KVM acivities while follower farmers though not attending the meetings but seeking knowledge and information from key and nodal farmers of the villages themselves.

Mode of Functioning

The group meets on 5th (including holiday) of every month for organizing a meaningful seminar-*cum*-discussion on the pre-decided subject and venue based on season and farmers need. A scientist, extension worker or an experienced farmer is selected as speaker of the topic. The lunch for the programme is also sponsored by an individual farmer or by a group of farmers or an institution. As a result, crop-wise group or enterprise-wise farmers groups are formed, viz., papaya, pomegranate, orange, cotton, chilli, pumpkin, cauliflower etc. The seminar usually starts in the forenoon and continues up to the evening. The seminar is started by oral quiz competition based on last month's seminar. The member giving correct answers of three consecutive questions is awarded with silver medal. Three such medals are awarded in each seminar. Members also voluntarily sponsor these medals. The seminar is followed by question and answer session and group interaction. The topic/venue for next month seminar or lecture is decided in the preceding month. Transport charges for attending the meeting are borne by participating farmers. Thc participants are provided with a pad and a pen on cost basis on their demand to note the important points of the seminar/lecture.

Role of KVK in Organizing Krishi Vigyan Mandal

Training Organizer of the KVK acts as Member Secretary of the Krishi Vigyan Mandal. Organizing seminars, group discussion, inviting scientists etc., is the responsibility of the KVK.

The year-wise details of the number of seminars conducted, topics covered and the number of farmers participated is given in Table 21.1. It is seen from the data that during the last six and half years total of 73 seminars covering 119 topics and participation of 10,383 farmers have been achieved.

Table 21.1 : Details of Seminars organized by KVM during 1996-97 to 2002-03

Sr. No.	*No. of Year**	*No. of Seminars*	*No of Topics*	*Farmers*
1.	1996-97	01	–	314
2.	1997-98	12	20	1643
3.	1998-99	12	14	1131
4.	1999-00	12	18	1288
5.	2000-01	12	24	1625
6.	2001-02	12	18	2540
7.	2002-03	12	25	1842
	Total	73	119	10383

*Reporting year considered as October-September

Some of the important topics covered are IPM in cotton, scientific irrigation methods or planting methods in sugarcane, Bahar management in Mosambi, Control of gummosis and dieback disease in oranges, Market information, WTO and its impact on Indian Agriculture etc.

Coverage of taluka (Block) - wise villages in the district is given in Table 21.2. It reveals that due to KVM, it was possible to cover more than 1/3 villages (31.74%) in the district in addition to 66 villages from out of district in a very short period of time. It was never possible to cover so many villages as per the KVK mandates.

Table 21.2 : No. of Taluka-wise villages covered by KVM during 1996-97 to 2002-03

Sr. No.	Name of Taluka	No. of Total Villages in Taluka	No. of Villages covered	Percentage
1.	Jalna	148	95	64.18
2.	Badnapur	93	45	48.38
3.	Jaffrabad	102	26	25.49
4.	Bhokardan	157	21	13.37
5.	Partur	98	27	27.55
6.	Mantha	117	21	17.94
7.	Ambad	135	34	25.18
8.	Ghansawangi	117	38	32.47
9.	Other districts	—	66	—
	Total	967	373	31. 74

Note : Other district villages are not included in the calculation of percentage.

Year-wise details of resource persons involved in KVM during 1997-98 to 2002-03 is given in Table 21.3. Table shows that total 139 resource persons are involved in KVM from different sources like SAUs, State Agricultural Departments, Private Officials, Innovative Farmers, ICAR, KVKs etc.

Table 21.3 : Year-wise details of Resource Person involved in KVM during 1997-98 to 2002-03

Sr. No.	Category Resource persons	No. of Participants						Total
		97-98	98-99	99-00	00-01	01-02	02-03	
1.	SAU	10	05	1	08	03	14	51
2.	State Agril. Departments	03	—	01	03	03	01	11
3.	Private Officials	03	06	03	05	09	03	29
4.	Innovative Farmers	02	01	02	10	02	05	22
5.	ICAR	—	—	—	—	02	01	03
6.	KVKs	—	01	—	—	—	01	04
7.	Others	02	04	03	06	04	—	19
	Total	20	17	22	32	23	25	139

Impact of the Programme

To know the effectiveness of Krishi Vigyan Mandal Zonal Coordination Unit, Hyderabad and KVK Jalna conducted an evalution study. One page proforma was developed in Marathi language and administrated to a group of farmers randomly. The sample consisted of 44 members. The findings of the study reveal that due to participation in KVM and group interactions Krishi Vigyan Mandal members undertaken number of activities. Group activities undertaken by KVM members is presented in Table 21.4.

Table 21.4 : Group Activities undertaken by Krishi Vigyan Mandal

S. No.	*Innovative technologies adopted by group*	*No. of KVM members adopting technology*	*Average yield/unit*	*Returns in terms of Rs./ Unit*	*Remarks*
1.	IPM in Cotton	75	15 q/ha	30,000	These members are from major 8 villages
2.	Group farming in summer chilli by using low cost "Pepsi tube" dripirrigation system	12	125 q/ha	1,25,000/ha	
3.	Group	10	150 q/ha	1,50,000/ha	farming in Bhendi
4.	Silkworm rearing	6	400 kg	54,800/ ha cocoon/ha	These members are from major 2 villages
5.	Group marketing of grapes, Pomegranates and sweet oranges	40	—	Grape Rs. 2000/t Rs.2500/t	Farmers got 20 to 25% higher rates than local market
6.	Aonla	5	—	—	Fruiting Cultivation is in progress Substitute Crop for Sweet Orange
7.	Accelerated composting technology by using bacterial compost	24	6 trollies from one bed of 40× 6′ ×3′	Rs. 3600/bed	Compost is not sold but used for their own farm

The data presented in Table 21.4 reveals that after becoming the KVM member farmers have adopted latest technologies, viz. IPM practices in cotton, low cost drip irrigation systems, use of bacterial culture for quick decomposition of Farm Yard Manure. Further group marketing of fruits, viz. grapes, pomegranate and sweet oranges gave 20-25 per cent higher returns than selling to the local traders. The above findings indicated that the farmer groups like Krishi Vigyan Mandal are effective in exchange of new ideas, adoption of new technologies and promotion of group marketing.

The findings are inconformity with the studies carried out by Colling, 1994; Mattee and Lassalle, 1994 ; Sperling, 1994; Carroll and Baitemmam, 1987; Garforth, 1994; Morss, 1976; Romenoff, 1993 (Ref. Arnaiz, 1995) that the farmers groups/organizations when engage in technology development and transfer activities it is usually in partnership, either with service NGOs or Government institutions.

Further, the opinion of the group members were sought about the effectiveness of Krishi Vigyan Mandal and they expressed that KVM is most useful in solving the problems of the farmers and created a platform for exchange of new technology, ideas, seeds, seedlings among the group members. The KVM felicitated in establishing rapport with experts, progressive farmers and scientists. It has empowered the progressive farmers further through selection as speakers for seminar topics. KVM members also got recognition at village and district level and received help from government officers.

REFERENCES

1. Kumar, K, (1987). Conducting Group Interviews in Developing Countries, AID Programme Design and Evaluation Methodology Report No. 8, USAID Washington.
2. Arnaiz, Maria E.O., (1995). Farmer's Organisations in the Technology Change Process (An annonated Bibilography) Network Paper 53 Odi January, London NW 1 4NS UK. pp. 2-3; 6.

Chapter - 22

Extension Approach for Sustainable Aquaculture in India

M. Kumaran

Introduction

Fisheries have emerged as a giant sector with an annual turnover of more than Rs.220 billion, which accounts for 1.4 per cent of the total GDP. Fisheries generate 5.7 million tonnes (2.83 million metric tonnes from marine and 2.82 million metric tonnes from inland. Doctors of valuable animal protein and feed at least 500 million people every year, meeting the human basic needs of nutrition towards food security and sustainable livelihood. Export earnings from fish and fishery products is 5957.05 million (2001-02). More than 6 million fishermen and fish farmers depend on fisheries and aquaculture for their livelihood. Shrimp accounts to about 28 per cent of the export (in terms of volume) and about 66 per cent in terms of value. A growth rate of 2.5 per cent been proposed for marine fisheries during the Tenth Five Year Plan (2002-07) and 8 per cent for inland water fishes and aquaculture development. The target for Tenth Five Year Plan regarding fisheries production is 8.09 MMT out of which 3.26 MMT contributed from marine sector and 4.83 MMT from inland sector. Since, the marine production has been stagnating over the years, inland aquaculture which contributes half of India' fish production utilizing only 16 per cent of its brackishwater and half of its freshwater culture potential should be the focus to produce more as envisaged in the Tenth Five Year Plan.

Aquaculture production could be multiplied through expansion of area, improvement in per capita productivity through adoption of Better Management practices, diversification of culture species and a vibrant extension machinery which create awareness, disseminate appropriate technology (Education), convince farming community to adopt such technologies (Motivation), Feedback and refinement technology to suit the farming situation (Assessment and Refinement) and act as a link between the research and user system (Liaison). The objective of aquaculture extension is to persuade and help aqua farmers and fishing communities to improve their socio-economic conditions and quality of life through their farming practices for hike in production and income. Efficient extension services are required to support the existing farmers and the new entrants for effectively promotion of equitable and sustainable development of aquaculture that will contribute to overall rural socio-economic life.

Existing Scenrio of Acquculture Extension

The organizations providing fisheries extension services in India are categorized and presented in Table 22.1. Fisheries Division of the Department of Agriculture and Cooperation under the Union Ministry of Agriculture is the nodal agency for co-coordinating fisheries development programmes in the country and in formulating major policy guidelines for the States of the country. Fisheries extension programmes are planned both at Central and State levels and implemented mainly through State Department of Fisheries (SDOF).

What Ails Aquaculture Extension?

Fisheries extension programmes, including aquaculture extension, are executed at field level exclusively by the respective State Fisheries Departments. The Fish Farmers Development Agency (FFDAs) and Brackish Water Fish Farmers Development Agency (BFDAs) at the district level is the important extension machinery for promotion of aquaculture. However, the FFDA/BFDA mechanism could not be supported with technology innovations, technical improvements due to its limited access to subject-matter specialists and lack of basic infrastructure for problem-solving at

Table 22.1 : Providers of Aquaculture Extension Services in India

Sl. No.	*Stream/Organization*	*Extension Programme/ Scheme*	*Activities*
1.	Fisheries Division, Ministry of Agriculture and Cooperation, Government of India	Aquaculture Authority of India Fish Farmers Development Agency (FFDA) and Brackishwater Fish Farmer Development Agency (BFDA)	Issuing Licence Information transfer Help the farmers from Project preparation to marketing Distribution of literature Technical advise Mobilizing water area on lease Training Field Demonstrations Organisation of seminars, workshops etc.
2.	State/UTs Departments of Fisheries	FFDA and BFDA	Issuing Licence Help the farmers from Project preparation to marketing Distribution of literature Technical advice Mobilising water area on lease Training Seed supply on cost basis Subsidy assistance Demonstrations Farmers meet, field days
3.	Indian Council of Agricultural Research (ICAR)	Krishi Vigyan Kendra (KVK), Trainers Training Centre (TTC) and TOT progress of Research Institutes of like Central Institute of Brackishwater Aquaculture (CIBA) and Central Institute of Fresh-water Aquaculture (CIFA)	Vocational training programes Seed supply on cost basis Demonstration Farmers meet, field days Technical advice Distribution of literature Awareness programme Exhibitions Institute Village Linkage Programme (IVLP) Organisation of seminars, workshops etc
4.	State Agricultural Universities (UAUs)- College of Fisheries	TOT Programmes as part of Education	Training Awareness programmes Distribution of extension materials Organisation of seminars, workshops etc.
5.	(Marine Products, Export Development Agency) Ministry of Commerce, Government of India	Export promotion through fresh and brackishwater aquaculture	Help the farmers for project preparation to marketing Technical service Subsidy assistance Training

(Contd.)

Table 22.1 *(Contd.)*

		Assistance in project preparation Assistance in obtaining credit Distribution of literature Assistance in marketing, exports, duty reduction, quarantine
6. Other Development Departments in the Centre and States (Ministry of Rural Development, DBT, DST, DAE etc.) Department of Bio-technology	R&D efforts to improve the technology package and pilot scale demonstration, rural programmes to improve the living standard of rural people	Welfare measures Demonstration in pilot scale Training Financial support for extension projects, printing literature
7. Non-Governmental Organisations and Private Agencies (Feed and other input companies, consultants, progressive farmers, inputs dealers, etc.)	Integrated Rural Development	Campaigns for eco-friendly aquaculture Technical advice Supply of inputs on credit basis Farmers' meet Farm trips Organising rural poor Mobilizing water bodies on lease Input assistance Group savings Balance development of coastal areas.

local level which has made them ineffective. The following are some of the constraints for the inefficient aquaculture extension services.

Unlike the agriculture extension system at the state level fisheries extension system does not have the primary role of extension and mobilising support services to the fish farmers. There is no formal well-established system for information delivery and to obtain feedback from the field.

— Inadequate skilled manpower to better the needs of the fish farmers.

— Major part of the reports meet the requirement of

budget monitoring and no qualitative information are quite essential for effective decision-making.

— Mismatch of trained personnel with their work and are centered towards welfare measures.

— Lack of necessary resources and infrastructure at field level.

— Activities under extension-training function are reported to concentrate only on quantitative information. Aspects such as need, adequacy, feedback and suggestions for improvements do not form part of these reports.

— There is no systematic interaction with research institutions, fisheries colleges and other technology generators.

— Small holdings, site-specific problems and erratic distribution of farms in distant locations.

Research-Extension-Farmer Linkage — An Extension Intervention

Effective technology transfer requires firm linkages between research and extension. Technology is developed through research and extension agents must have access to researchers to know and understand the latest technology to ensure its successful transmission to farmers. At the same time, a good feedback from the farming community obtained through extension personnel would be of great help to the researchers in planning relevant research programmes or in modifying the existing programmes in tune with the farmer needs. At present, the linkages between the technology generators and adopters are quite weak. There is a very limited interaction between the Fisheries Colleges of State Agricultural Universities and the Fisheries Research Institutes of the ICAR even when they are located in proximity. Hence, each other are oblivious of what the other is doing. The colleges of fisheries have to play a crucial role in coordinating research and extension functions so that appropriate technology could reach the end user.

The farming community has a closer contact with State Department of Fisheries (SDF) and Marine Products Export Development Authority (MPEDA) than the other agencies. This may be due to the reason that MPEDA is the promoter of shrimp farming who provide subsidies and training in the initial stages of commercial shrimp farming. The Assistant Director of Fisheries (Aquaculture) of the State Department of Fisheries is the nodal officer to recommend for subsides/insurance provided by the government and banks and to arrange licence issued by the Aquaculture Authority. Studies of AA 2001; Kumar, 1996 and Kumaran, 2001/2003 infers that at present the linkage and coordination among the fisheries institutions are negligible and the major technical players as far as aquaculture is concerned are input dealers and technicians. It proves that the linkage between the private extension sources and the farming community is very strong since both depend on each other.

Inputs dealers and their retailers/representives are the most important sources of information for all technical matters like seed and stocking, feed and supplementary feeding, water quality management, health, harvest, marketing etc. The input dealers/traders employ biology graduates as field technicians/assistant who had undergone training in shrimp farming. These field technicians regularly visit their customer shrimp farms to provide all necessary technical assistance. Input companies and/or dealer do not have any contact either with Research Institutions/Fisheries Colleges nor with the extension and promotional agencies.

Suggested Extension Approach

Several experts (BOBP, 2000; Singh and Sampth, 1981, 1982, 1983 and 1990; Radheshyam and Kumar, 1982; Tripathy *et al.*, 1982; Rahiman *et al.*, 1991, Alagarswami, 1995; Krishna 1995 & 2000; and Kumaran, 2001 & 2003) emphasized that a socio-environ friendly sustainable fish production is possible provided there should be a closer linkage and coordination among the main players of fisheries. There should be workshops, discussion meetings and interactions before and after completion of one cropping cycle nor with the extension and promotional agencies.

Suggested Extension Approach

Several experts (BOBP, 2000; Singh and Sampth, 1981, 1982, 1983 and 1990; Radheshyam and Kumar, 1982; Tripathy *et al.*, 1982; Rahiman *et al.*, 1991, Alagarswami, 1995; Krishna 1995 & 2000; and Kumaran, 2001 & 2003) emphasized that a socio-environ friendly sustainable fish production is possible provided there should be a closer linkage and coordination among the main players of fisheries. There should be workshops, discussion meetings and interactions before and after completion of one cropping cycle among research and extension and other stakeholders (research, extension, promotional agencies, public, NGOs, and farming community). These sessions provide a common platform for the extension agents, senior fisheries officers, fish farmers and scientists of research institutions, fisheries colleges, other departments working in fisheries, delivering inputs and export agencies to meet and discuss various emerging issues. The workshop also provides opportunity for the participants to get first hand information about the performance of a particular technology packages at farm level.

The technology generates will also get an opportunity to brief the group about some of the recent findings and technologies developed by the research institutions and provide ready-made solutions to some of the emerging technical problems. Scientists can also validate their findings in farmers ponds to understand the field situation a discussions and exchange of ideas help scientists to get an insight into field problems and plan cost-effective field research. The research institutes, fisheries colleges and extension agencies should sensitize the politicians, revenue and other development departments and the coastal public that fisheries/ aquaculture is not an illegal activity harmful to any enterprise or anybody. In addition the following aspects are also suggested:

- Reorganization and reorientation of extension machinery in State Fisheries Departments and leaving away the

(SDOF, MPEDA) to get their products certified by them and involved in marketing the products developed by the researchers.

Integration of Participatory and Group Approach for Technology Transfer towards Sustainable Fisheries

Based on the studies conducted, experiences gathered and discussion with experts associated with fisheries the author suggests an integration of 'Participatory-cum-group approach for Technology Transfer towards Sustainable Fisheries'. This model suggests a two-way symbiotic link among all the stakeholders of fisheries. It proposes a regional workshop twice in a year for two days one before the first season and the other in the post first season or pre-second season at regional level with the presence of State Fisheries Minister/Director of Fisheries of the State. Resource persons from Research Institutes and Fisheries College (RIS & FC), JDs, ADs, SMSs, FEOs, Reps. Of Farmers, hatcheries, feed and other inputs, NGOs etc. will participate in the workshop. During first day the performance in the concluded season, technical and administrative constraints, area development, new issues etc. are to be discussed extensively and sorted out. Representatives of farming community, inputs dealers, exporters, coastal public, fishers, NGOs, would also present in the meeting and express their views and get their grievances discussed and addressed with each other. In the second day the resource persons from the RIS and FCs will explain the technical messages and finalization of technical messages for the forthcoming season. The same should be printed and distributed in advance so that the necessary clarifications could be obtained during the discussion.

District level workshops at quarterly intervals should be held for a day in the presence of the Joint Director of Fisheries in the region in which resource persons of RIs & FC, AA, ADs, SMSs, FEO, Reps. of feed and other inputs, NGOs etc. would participate. The messages finalized for the season in the regional workshops and the problems of local importance should be discussed thoroughly. At Taluk/Block level Assistant Directors of Fisheries

Extension Personnels, and private extension agents working with fishers/farmers would present to discuss and plan the monthly programmes. The lower level extension personnel and private extension agents are trained on the messages and skills to be delivered the end users. At field level the same may be communicated to the Farmers Group Representatives and they in turn disseminate the same to the fellow farmers. Similar way the field problems will be communicated to the research and administration organically at regular intervals. Two factors vital for successful technology development and transfer are integration of researcher, extension worker, other contributors and farmers' and the other is strong farmers' organization at the grassroots level.

Conclusion

1. Fisheries extension is a neglected field in the overall fisheries development programmes so far in India. Extension should be equally strong on par with technology development in fisheries and so far no effort has been made to make it strong. Farmers have established a confidence with private extension group.
2. Public extension system is facing a problem of manpower and budget. Hence, linking institutional and non-institutional extension sources is the appropriate measure in this present context.
3. Identification, registration and capacity building of consultants, farm technicians, inputs dealers and farm leaders will help in delivering relevant scientific information to the farming community.
4. Every sub-system is functioning in isolation and there is no meaningful interaction so far to such an extent to evolve a system, which is useful for the farming community in increasing their pond productivity ultimately to increase the overall fisheries production.
5. Technology evolvers, private and public extension agents, and farmers should work together as a team for any successful aquacultural development programme and

extension system requires extension agents who are more confident of production techniques and in extension methods.

6. The extension system needs to be redefined, redesigned and equipped with several solutions to the problems of different groups of farmers with varied dimensions of land holdings, farming situations and income.

REFERENCES

Alagarswami, K.(1995). 'Strategies for sustainable shrimp farming and transfer of technology', *Proc. of the N. Workshop on Transfer of Technology for Sustainable Shrimp Farming*, CIBA, Chennai, pp. 29-42.

Aquaculture Authority (AA). (2001). 'Shrimp Aquaculture and the Environment –An Environment Impact Assessment Report' submitted to the Supreme Court of India, Government of India, Chennai, p. 92.

Bay of Bengal Programme (BOBP). (2000). Monitoring and Evaluation and Management information systems of Department of Fisheries, Govt. of Tamil Nadu, India, BOBP/MM/ 04, p. 62.

Krishna, S. (1995). Improving Transfer of Technology and Extension Systems, *Proc. of the N. Workshop on Transfer of Technology for Sustainable Shrimp Farming*, CIBA, Chennai, pp.146-53.

Krishna, S., (2000) Experiences in Aquaculture: Some Lessons for Extension: in Krishnan and Birthal (Ed.) *Proc. Aquaculture Development in India: Problems and Prospects,* Series no.7, NCAP, New Delhi, pp. 87-97.

Kumar, D., (1996). Aquaculture Extension Services Review: India, FAO Fisheries Circular, No. 906, FAO, UN, Rome, p. 71.

Kumar, D., (1999). *Trickle Down System (TDS) of Aquaculture Extension for Rural Development*, RAP Publication 23, FAO, UN, Bangkok.

Kumaran, M., K. Ponnusamy and N. Kalaimani. (2001). 'Diffusion and Adoption of Shrimp Farming Technologies' Paper presented in the International Symposium on Fish for Nutritional Security in the 21st Century, at CIFE, Mumbai, 4-6th December, 2001.

Kumaran, M., Chandrasekaran, V.S. and Kalaimani N. (2003). 'Aquaculture Extension— The Neglected Aspect in Aquaculture Development', *Fishing Chimes,* 23(6) : pp. 29-34.

Monu, E.D. (1995). Technology development and dissemination in agriculture: a critique of the dominant model. *Development Africa.*20 (2): pp. 21-39.

Neelameghan, A. (1973). Technology Transfer : A System Analysis Approach to the Study of Its Features. *Annals-of-Library-Science-and-Documentation,* 20(1-4), pp. 1-25.

Ponnusamy, K., Gopinathan, K. and M. Krishnan. (1999). A Study on Cultural and Economic Aspects of Aquacultural Farming Practices and Extension needs of Farmers in Prakasam District of Andhra Pradesh, Paper presented in National Seminar on Development and Transfer of Fisheries Technology, FC&RI, Thoothukudi, Tamil Nadu, February. 3-5.

Radheshyam and D. Kumar, (1982). Fisheries Extension for Rural Agriculture. In Souvenir. Workshop on the Development of Inland Fisheries in Orissa through Institutional Financing. FFDA. Balasore, Orissa, India. 6-8 March. pp. 99-105.

Rahiman, O.A., A.M. Tampi, and C. Bhaskaran. (1991). Transfer of Technology System : Functional linkages between and among four sub-systems. *Journal-of Extension-Systems*, 7:1, pp. 91-96.

Singh, V.D. and V. Sampath. (1981). Introduction of Fisheries Extension Service for Development of Inland Fish Farming in India. *Pub. Fish. Bull.* pp. 29-34.

Singh, V.D. and V. Sampath. (1982). Role of Fisheries Extension in Rural Areas with particular reference to Uttar Pradesh, *Indian Farmers Digest*, XV (1): pp. 41-43.

Singh, V.D. and V. Sampath. (1990). A Well Organised Fisheries Extension Service System a Prerequisite for Development of Agriculture for Higher Production. *Pub. Fish. Bull.* Xiv(2): pp.1-9.

Tripathy, N.K., D. Kumar. (1982). Concept of Community Fish Framing. In Souvenir, Workshop on the Development of Inland Fisheries in Orissa Through Institutional Finance, Balsore, India, p. 23.

Yadava, Y.S. (2002). 'Shrimp Farming in India—Lessons and Challenges in Sustainable Development' *Aquaculture Authority News*, 1(1): pp.1-4.

Index